Practical
Geometry Part Two

Kwang S. Ko, Ph.D.
Alexander H. Ko

1. *A SELF-STUDY GUIDE*
2. *PRACTICE PROBLEMS*
3. *SELF-TESTS*
4. *REVIEW TEST*
5. *FULL ANSWER KEY*

Copyright © 2015 by Kwangseuk Ko & Alexander Ko

All rights reserved.
No part of this work may be reproduced or transmitted in any form, by any means, or by any information storage or retrieval system, without the prior written permission of the copyright owner.

Request for information should be addressed to 7722 Camino Noguera, San Diego, CA 92122
Visit our website at www.iqmaths.com

ISBN: 978-1523362011

Printed in the United States of America

10 9 8 7 6 5 4 3 2

Part Two

CONTENTS

Chapter 7	**SIMILARITY**	1
	Concepts, Examples, Formulas, and Vocabulary	1
	Practice	12
	Answer to Practice	26
	Self-Test	27
	Answer to Self-Test	34
Chapter 8	**RIGHT TRIANGLES**	35
	Concepts, Examples, Formulas, and Vocabulary	35
	Practice	46
	Answer to Practice	60
	Self-Test	61
	Answer to Self-Test	68
Chapter 9	**CIRCLES**	69
	Concepts, Examples, Formulas, and Vocabulary	69
	Practice	80
	Answer to Practice	94
	Self-Test	95
	Answer to Self-Test	105
Chapter 10	**AREA OF POLYGONS AND CIRCLES**	106
	Concepts, Examples, Formulas, and Vocabulary	106
	Practice	114
	Answer to Practice	125
	Self-Test	126
	Answer to Self-Test	134
Chapter 11	**SURFACE AREA AND VOLUME OF SOLIDS**	135
	Concepts, Examples, Formulas, and Vocabulary	135
	Practice	149
	Answer to Practice	164
	Self-Test	166
	Answer to Self-Test	174

REVIEW TEST II — 175
    Chapter 7 – Chapter 11 — 175
    Answer to Review Test II — 186

APPENDIX — 187

Part One*

*The following is a separate book (Practical Geometry –part one).

CONTENTS

Chapter 1 **BASIC GEOMETRY**

Chapter 2 **INDUCTIVE AND DEDUCTIVE REASONING**

Chapter 3 **PERPENDICULAR AND PARALLEL LINES**

Chapter 4 **CONGRUENT TRIANGLES**

Chapter 5 **TRIANGLES**

Chapter 6 **QUADRILATERALS**

REVIEW TEST I

APPENDIX

CHAPTER 7
Similarity

In this chapter, you will write and simplify the ratio of two numbers, discern proportions and their properties, identify the properties of similar polygons, and prove that two triangles can be similar with different postulates and theorems, as well as learning how to identify dilations and its properties. Moreover, you will learn how to similar polygons can relate to proportions and their connections to the real world.

CONCEPTS   EXAMPLES   FORMULAS   VOCABULARY

7-1. Express as a ratio.

   a. 1 to 4   b. 4 feet to 7 feet

   SOLUTION

   A ratio is a comparison of two numbers by division. It can be written out by using three methods as shown by the following: 1) using the phrase " a to b", 2) using a colon "a : b", or writing it out as a fraction "$\frac{a}{b}$".
   a. "1 to 4" means $1 : 4 = \frac{1}{4}$
   b. "4 feet to 7 feet" means $4 : 7 = \frac{4}{7}$

   * Ratio: a ratio is a comparison of two numbers by division.
   * Equivalent ratios: two ratios that reduced to the same number.
   * Proportion: a proportion is an equation that states that two ratios or fractions are the same.

7-2. Express as a ratio.

   a. 4 g to 5 kg.   b. 13 ft to 20 in.

   SOLUTION

   When you compare two measurements in a ratio, they should both have the same unit. Otherwise, if they don't match each other, they will have to be converted to the same unit.
   a. "4 g to 5 kg" = 4 g to 5 kg × (1000g/1 kg)   Convert to like units.
                 = 4 g to 5000g
                 = $\frac{4}{4}$ g : $\frac{5000}{4}$ g = 1g : 1250 g   Simplify.

   b. "13 ft to 20 in." = 13 ft × (12 in./1 ft) to 20 in.   Convert to like units.
                    = 156 in to 20 in.
                    = 156 : 20 = $\frac{156}{4} : \frac{20}{4}$ Simplify.
                    = 39 in : 5 in

7-3. Write the ratios in the statement as an equation and determine whether if they are a proportion or not. Explain.

a. 3 to 8 as 9 to 24.   b. 5 to 7 as 10 to 11.

SOLUTION

A proportion is an equation that equates two ratios.

a. "3 to 8" means $3:8 = \frac{3}{8}$ and "9 to 24" means $9:24 = \frac{9}{24}$

So you can determine whether the statement is a proportion if the cross products of the ration are equal to each other.

$$\frac{3}{8} \times \frac{9}{24}$$

$(3)(24) = (9)(8)$ Cross Products Property.
$72 = 72$ Simplify.

The cross products are equal to each other. Therefore the statement is a proportion and the equation is $\frac{3}{8} = \frac{9}{24}$.

b. "5 to 7" means $5:7$ or $\frac{5}{7}$ and "10 to 11" means $10:11$ or $\frac{10}{11}$.

So you can determine whether the statement is a proportion if the cross products are equal to each other.

$$\frac{5}{7} \times \frac{10}{11}$$

$(5)(11) = (10)(7)$ Cross Products Property
$70 \neq 55$ Simplify.

The cross products are NOT equal to each other. So, the statement is NOT a proportion and the equation is $\frac{5}{7} \neq \frac{10}{11}$.

7-4. Simplify the ratio.

$$\frac{10}{x} = \frac{2}{7}$$

SOLUTION

You can solve this problem in two ways. You can use the Cross Product Property to solve the equation.

$\frac{10}{x} = \frac{2}{7}$ Original proportion
$(10)(7) = 2x$ Cross Product Property
$70 = 2x$ Simplify.
$x = 35$ Divide both sides by 2.

Or you can use the Reciprocal Property as an alternative to solving the equation.

Second, use reciprocal property to solve the equation.

$\frac{10}{x} = \frac{2}{7}$ Original proportion

$\frac{x}{10} = \frac{7}{2}$ Reciprocal property

$x = (10)\frac{7}{2}$ Multiply both sides by 10.

$x = 35$ Simplify.

So the value of x is 35.

7-5. Complete each sentence.

a. If $\frac{3}{x} = \frac{5}{y}$, then $\frac{3}{5} =$

b. If $\frac{x}{9} = \frac{7}{y}$, then $\frac{x+9}{9} =$

SOLUTION

There are two additional properties of proportions as following: 1) if $\frac{a}{b} = \frac{c}{d}$, then $\frac{a}{c} = \frac{b}{d}$, where $b = 0$ and $d = 0$, then $ad = bc$ and 2) if $\frac{a}{b} = \frac{c}{d}$, then $\frac{a+b}{b} = \frac{c+d}{d}$.

a. $\frac{3}{x} = \frac{5}{y}$

$\frac{3}{5} = \frac{x}{y}$

b. $\frac{x}{9} = \frac{7}{y}$

$\frac{x+9}{9} = \frac{7+y}{y}$

7-6. In BC ∥ DE, find the value of x.

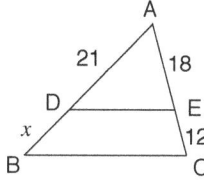

SOLUTION

You can apply the two properties of proportions.

$\frac{AD}{AB} = \frac{AE}{AC}$ Write proportion.

$\frac{21}{x+21} = \frac{18}{30}$ Substitute.

$18(x + 21) = (21)(30)$ Cross Product Property

$18x + 378 = 630$ Distribute.

$18x = 252$ Subtract.

$x = 14$ Divide both sides by 18.

So the value of x is 14.

7-7. Find the geometric mean of the two numbers.

    a. 4 and 18   b. $7x$ and 10

SOLUTION

a. The geometric mean of 4 and 18 is the positive number x.

$\dfrac{4}{x} = \dfrac{x}{18}$ Set up proportion

$x^2 = (4)(18)$ Cross Product Property

$x = \sqrt{(4)(18)}$ Simplify.

$x = 6\sqrt{2}$ Simplify.

So the value of x is $6\sqrt{2}$.

b. The geometric mean of $7x$ and 10 is the positive number x.

$\dfrac{7x}{x} = \dfrac{x}{10}$ Set up proportion.

$x^2 = (7x)(10)$ Cross Product Property

$x = 70$ Simplify.

So the value of x is 70.

* To find the geometric mean.

    means   means

$4 : x = x : 18 \;\Rightarrow\; x^2 = (4)(18)$ $7x : x = x : 10 \;\Rightarrow\; x^2 = (7x)(10)$

  extremes   extremes

To find the geometric means of three numbers such as 3, 7, and 8, insert them in an equation as displayed: $x^3 = (3)(7)(8)$, $x = \sqrt[3]{(3)(7)(8)}$, and solve for x. Just remember that a set of n numbers will take the nth root.

7-8. Use the given information to find the lengths of AB and BD.

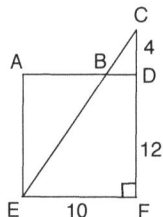

Chapter 7 Similarity

SOLUTION

You can draw the diagrams separately as △BCD and △ECF. The two triangles are similar. And the length of CF is the sum of the lengths of CD and DF.

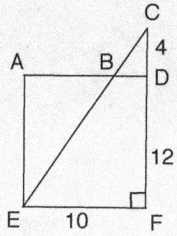 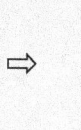 ⇒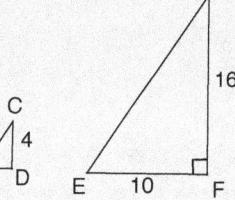

$\dfrac{CD}{CF} = \dfrac{BD}{EF}$ Write proportion.

$\dfrac{4}{16} = \dfrac{BD}{10}$ Substitute.

$4(10) = (BD)(16)$ Cross Product Property
$40 = BD(16)$ Multiply.
$2.5 = BD$ Divide.

$AD = AB + BD$ Segment Addition Postulate.
$AB = AD - BD$ Find the length of AB/ Subtract both sides by BD.
$AB = 10 - 2.5$ Substitute.
$AB = 7.5$ Subtract and answer.

7-9. In the figure below, the polygons shown are similar. Write a similarity statement that includes the pairs of congruent angles and corresponding sides.

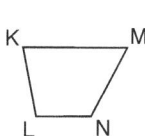 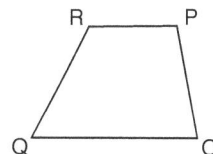

SOLUTION

The corresponding angles of two similar polygons are always congruent. Show that the lengths of corresponding sides are proportional.
In the diagram, KLMN ~ OPQR. ∠K, ∠O, ∠L, ∠P, ∠N, ∠R, ∠M, ∠Q,

and $\dfrac{KM}{OQ} = \dfrac{KL}{OP} = \dfrac{MN}{QR} = \dfrac{LN}{PR}$

7-10. Tell whether the triangles are similar. Explain why or why not.

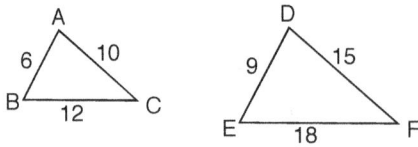

> **SOLUTION**
>
> △ABC and △DEF are similar if the corresponding sides are proportional.
>
> $\dfrac{AB}{DE} = \dfrac{6}{9} = \dfrac{2}{3}$ $\dfrac{BC}{EF} = \dfrac{12}{18} = \dfrac{2}{3}$ $\dfrac{AC}{DF} = \dfrac{10}{15} = \dfrac{2}{3}$
>
> Therefore, the two triangles are similar because of their corresponding sides are proportional.

7-11. In the diagram below, pentagons ABCDE and FGHIJ are similar. Find the value of x.

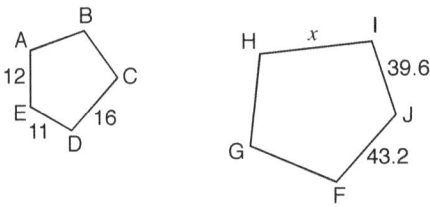

> **SOLUTION**
>
> You can find the scale factor for two or more similar polygons. The scale factor is a proportion that applies for all corresponding sides in similar polygons.
>
> $\dfrac{JF}{AE} = \dfrac{43.2}{12} = 3.6$ $\dfrac{IJ}{DE} = \dfrac{39.6}{11} = 3.6$
>
> Therefore, the scale factor is 3.6. Now you can use the scale factor in order to find the value of x.
>
> $\dfrac{HI}{CD} = \dfrac{x}{16} = 3.6$
>
> $x = (3.6)(16) = 57.6$. So, the length of IH is 57.6.

7-12. △ABC is similar to △DEF. The similarity ratio of △ABC to △DEF is 6: 4. What is the length of DE given that the length of AB is 18 cm?

> **SOLUTION**
>
> △ABC is similar to △DEF. Because DE is corresponds to AB, you can use the scale factor in order to find the length of DE. So given the information that the scale factor is $\frac{4}{6}$ or $\frac{2}{3}$ by using the scale factor, the length of DE is $\frac{2}{3}$ times the length of AB.
>
> $$DE = (\tfrac{2}{3})(18) = 12$$
>
> The length of DE is 12 cm.

7-13. Determine whether △ADE and △ABC are similar.

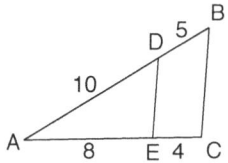

> **SOLUTION**
>
> Prove that △ABC and △ADE are similar by using the statement of proportionality so that
> $\frac{AD}{AB} = \frac{AE}{AC} = \frac{DE}{BC}$.
>
> $$\frac{AD}{AB} = \frac{10}{15} = \frac{2}{3}$$
>
> $$\frac{AE}{AC} = \frac{8}{12} = \frac{2}{3}$$
>
> So, △ABC and △ADE are similar because the ratios of their corresponding sides are equal.

7-14. △ABC is similar to △DEF. Find $m\angle A$ and $m\angle F$ for (a) and $m\angle B$ and $m\angle D$ for (b).

a. 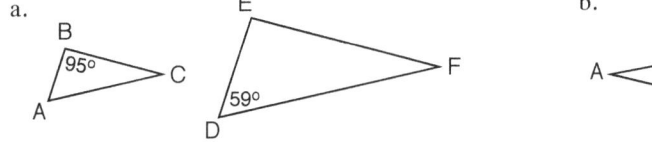 b.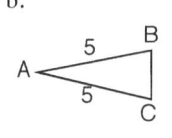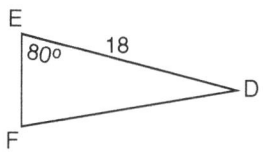

> **SOLUTION**
>
> a) By the Angle-Angle (AA) Similarity Postulate, if ∠B ≅ ∠E and ∠A ≅ ∠D, then △ABC and △DEF are similar. Because two triangles are similar, $m\angle A = m\angle D = 59°$ and $m\angle B = m\angle E = 95°$. By definition of a triangle, $m\angle D + m\angle E + m\angle F = 180°$. $59° + 95° + m\angle F = 180°$. So $m\angle F = 26°$.

b) ΔABC and ΔDEF are isosceles triangles because two of their side lengths are congruent. So by the Base Angles Theorem, which states that if two sides of a triangle are congruent, then the angles opposite these sides are congruent. ∠B is congruent to ∠C and ∠E is congruent to ∠F, so ∠B ≅ ∠C and ∠E ≅ ∠F. Also by the AA Similarity Postulate, $m\angle B = m\angle E = 80°$ and $m\angle E = m\angle F = 80°$. By the definition of a triangle, $m\angle D + m\angle E + m\angle F = 180°$. $m\angle D + 80° + 80° = 180°$. So $m\angle D = 20°$.

7-15. ΔABC is similar to ΔDEF below. Find the perimeter of ΔDEF if the perimeter of ΔABC is 32 cm.

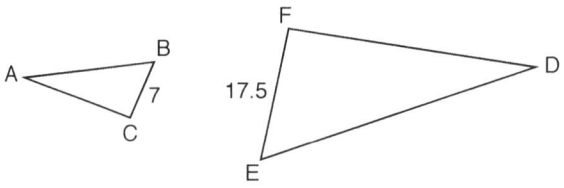

SOLUTION

You can use the scale factor to find the perimeter of ΔDEF. Since ΔABC and ΔDEF are similar, the perimeters of the two figures must have the same ratio.

$$\frac{EF}{BC} = \frac{17.5}{7} = 2.5$$

Therefore, the scale factor is 2.5. The perimeter of ΔDEF is (2.5)(32) = 80 cm.

7-16. Prove that ΔABC and ΔADE are similar.

Given: $\overline{DE} \parallel \overline{BC}$
Prove: ΔABC ~ ΔADE

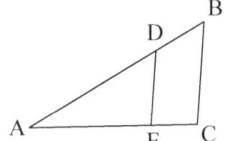

SOLUTION

Statements	Reasons
1. $\overline{DE} \parallel \overline{BC}$	1. Given
2. ∠E ≅ ∠C	2. Corresponding Angles Postulate
3. ∠A ≅ ∠A	3. Reflexive prop. of Congruence
4. ΔABC ≅ ΔADE	4. AA Congruence Postulate

7-17. Find the length of AD given that DE is parallel to BC.

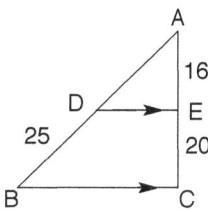

SOLUTION

By the Triangle Proportionality Theorem, if a line is parallel to one side of a triangle and intersects the midpoint of the other two sides (BC), then the line divides the two sides proportionally.

DE ∥ BC means $\frac{AD}{BD} = \frac{AE}{CE}$.

$\frac{AD}{25} = \frac{16}{20}$ Substitute.

AD(20) = (16)(25)   Cross Product Property

AD = 20   Simplify.

So, the length of AD is 20.

7-18. Is BD parallel to CD?

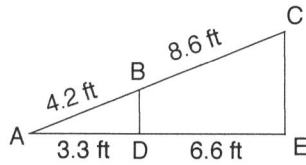

SOLUTION

You can apply the Converse of the Triangle Proportionality Theorem and show BD is parallel to CE if the sides are proven to be proportional.

$\frac{AD}{DE} = \frac{AB}{BC}$ means BD ∥ CE.

If the ratio of AB to BC is equal to the ratio of AD to DE, then BD is parallel to CE.

$\frac{4.2}{8.6} \stackrel{?}{=} \frac{3.3}{6.6}$ Substitute.

$(4.2)(6.6) \stackrel{?}{=} (3.3)(8.6)$ Cross Product Property

27.72 ≠ 28.38   Simplify.

Therefore, BD is not parallel to CE.

7-19. Find the value of *x*.

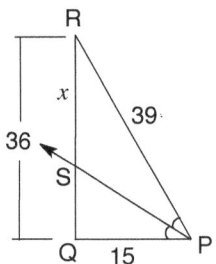

SOLUTION

By the Proportionality Theorem, if a ray ($\vec{PS}$) bisects an angle (∠RPQ) of a triangle, then it divides the opposite side into segments that are proportional to the lengths of the other two legs of the triangle. ($\frac{PQ}{PR} = \frac{QS}{RS}$). $\vec{PS}$ is an angle bisector of ∠RPQ.

$\frac{PQ}{PR} = \frac{QS}{RS}$	Triangle Proportionality Theorem
$\frac{15}{39} = \frac{36 - x}{x}$	Substitute.
$(15)(x) = (39)(36 - x)$	Cross Product Property
$15x = 1404 - 39x$	Distributive Property
$54x = 1404$	Add $39x$ to both sides.
$x = 26$	Divide both sides by 54.

So the value of *x* is 26.

7-20. Use the diagram, AB ∥ CD and CD ∥ EF. Find the length of CE.

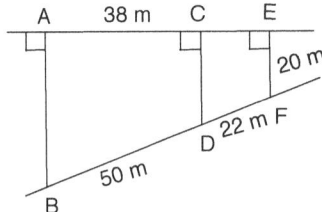

SOLUTION

By the definition of a transversal, if three parallel lines intersect two transversals, then they divide the transversals proportionally. So because the corresponding angles of ∠A, ∠C, and ∠E are congruent, the lines AB, CD, and EF are parallel.

Chapter 7   Similarity

$$\frac{AC}{CE} = \frac{BD}{DF}$$ Definition of proportions

$$\frac{38}{CE} = \frac{50}{22}$$ Substitute.

$(38)(22) = (50)(CE)$ Cross Product Property
$16.72 = CE$ Divide both sides by 50.

So the length of CE is 16.72.

7-21. Find the image of the point by using a dilation with the origin (0, 0) as the center and the scale factor given as $\frac{1}{5}$.

$$Q(5, 15)$$

SOLUTION

For a dilation with center C and scale factor k, you can find the image of a point by multiplying each coordinate by $k(x, y) = (kx, ky)$. So a dilation with the center (0, 0) is $\frac{1}{5}$ (5, 15) or (1, 3).

* Dilation: Dilation is an image that produced by enlarging or reducing the size, but not its shape.

* Name the Units:

Lengths	Volume
1 yard (yd.) = 3 feet (ft) = 36 inches (in.)	1 liter (L) = 10 deciliters (dL)
1 kilometer (km) = 1000 meter (m)	1 liter (L) = 100 centiliters (cL)
1 mile (mi.) = 1,760 yards (yd.) = 5,280 feet (ft)	1 liter (L) = 1,000 milliliters (mL)
Volume	Mass (weights)
1 pint = 2 cups (C)	1 kilogram (kg) = 1,000 grams (g)
1 quart = 2 pints (pt)	1 kilogram (kg) = 2.204 pounds
1 gallon = 4 quarts (qt)	1 pound (Ib) = 16 ounces (oz)
1 cup = 8 fluid ounces (fl oz)	

PRACTICES

1. In Exercises **a** and **b**, write the ratios in the statement as an equation and determine whether it is a proportion. Explain.
   ≪See Examples 7-1 to 7-3≫

   a. 7 is to 8 as 28 is to 33.

   b. 6 is to 10 as 4.2 is to 7.0.

2. In Exercises **a-d**, write the ratios in the statement as an equation and solve the proportions.
   ≪See Examples 7-1 to 7-3≫

   a. 11 to 48 as x to 22.

   b. $x + 3$ to 12 as $2x$ to 6.

   c. $(3x + 2)$ to 6 as 2 to 10.

   d. $(2x + 5)$ to 5 as 15 to 30.

3. In Exercises **a** and **b**, find the unknown lengths if the ratios of each side lengths in the shape are in the extended ratio of 1: 8.
   ≪See Example 7-6≫

   a.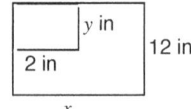

   b.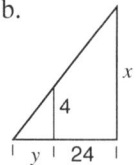

4. Use the graph for Exercises **a-c**, write the slope of a line as the ratio of the coordinate plane. Then simplify the ratio.
   ≪See Example 7-5≫

   a. $\overleftrightarrow{ST}$

   b. $\overleftrightarrow{MN}$

   c. $\overleftrightarrow{PQ}$

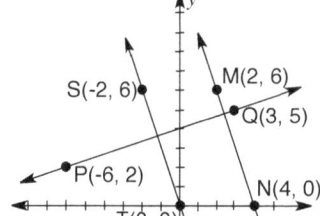

Chapter 7 Similarity

5. For Exercises **a-c**, write a ratio for each side length.
≪See Example 7-4≫

a.

b.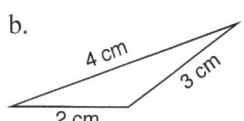

c.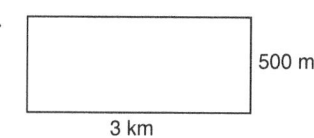

6. In Exercises **a-l**, simplify the ratio.
≪See Example 7-2≫

a. $\dfrac{10 \text{ yd}}{24 \text{ ft}}$ b. $\dfrac{640 \text{ g}}{0.32 \text{ kg}}$ c. $\dfrac{12 \text{ m}}{360 \text{ cm}}$

d. $\dfrac{2.5 \text{ m}}{500 \text{ cm}}$ e. $\dfrac{1.62 \text{ km}}{3 \text{ mi}}$ f. $\dfrac{45 \text{ ft}}{15 \text{ ft}}$

g. $\dfrac{16 \text{ oz}}{1 \text{ lb}}$ h. $\dfrac{4 \text{ lb}}{60 \text{ oz}}$ i. $\dfrac{72 \text{ cm}}{1.2 \text{ m}}$

j. $\dfrac{20 \text{ cm}}{1.5 \text{ m}}$ k. $\dfrac{4 \text{ ft}}{36 \text{ in}}$ l. $\dfrac{16 \text{ in}}{4 \text{ ft}}$

7. In Exercises **a-l**, solve the proportion.
≪See Example 7-5≫

a. $\dfrac{a}{12} = \dfrac{6}{10}$ b. $\dfrac{a}{16} = \dfrac{18}{22}$ c. $\dfrac{2y}{16} = \dfrac{y+3}{11}$

d. $\dfrac{2}{x} = \dfrac{1}{7}$ e. $\dfrac{y}{5} = \dfrac{4}{25}$ f. $\dfrac{13}{6} = \dfrac{m}{4}$

g. $\dfrac{2x-3}{7} = \dfrac{3}{5}$ h. $\dfrac{m}{3} = \dfrac{7}{27}$ i. $\dfrac{y+1}{3} = \dfrac{1}{6}$

j. $\dfrac{2(4x+7)}{9} = \dfrac{x}{7}$ k. $\dfrac{2}{3x+4} = \dfrac{2}{9}$ l. $\dfrac{2}{3} = \dfrac{7}{y}$

8. For Exercises **a-h**, write a proportion to solve each question.
≪See Examples 7-4 and 7-5≫

a. 4 notebooks for $5
12 notebooks for $y

b. 2 water bottles for $2.50
8 water bottles for $x

c. $4 for 2 pounds of apple
$y for 8 pounds of apple

d. $6.50 for 8-pack AA battery
$z for 24-pack AA battery

e. $18 for 3 pounds of T-bone steak
$x for 13 pounds of T-bone steak

f. 2 movie tickets for $15
7 movie tickets for $y

g. 105 miles traveled in 1 hours and 30 minutes
350 miles traveled in x hours

h. 12 mm raining per 1 hour
80 mm raining per x hour

9. For Exercises **a-f**, complete each proportion.
≪See Example 7-5≫

a. If $\frac{r}{12} = \frac{p}{4}$, then $\frac{r}{p} =$

b. If $\frac{p}{16} = \frac{q}{12}$, then $\frac{p}{q} =$

c. If $\frac{19}{s} = \frac{2}{t}$, then $\frac{19}{2} =$

d. If $\frac{27}{s} = \frac{8}{t}$, then $\frac{27}{8} =$

e. If $\frac{r}{45} = \frac{p}{5}$, then $\frac{r+45}{45} =$

f. If $\frac{2}{s} = \frac{6}{t}$, then $\frac{2+s}{s} =$

10. For Exercises **a-f**, find the values of x and y.
≪See Example 7-6≫

a.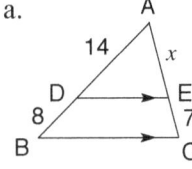

b.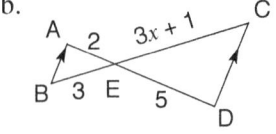

c.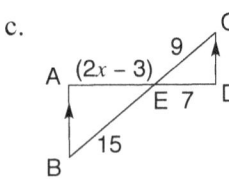

d.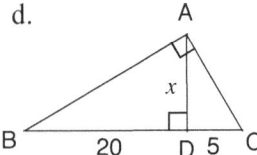

e.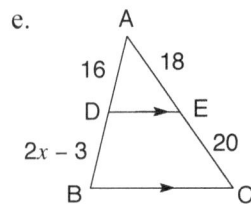

f.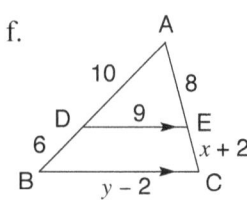

Chapter 7   Similarity   15

11. Find the geometric mean between each pair of numbers.
≪See Example 7-7≫

a. 25 and 40   b. 15 and 145   c. 4 and 121

d. 9c and 19c   e. 16c and 25c   f. 4 and 41

g. 4(x − 5) and 25(x + 5)   h. 9y and 45y   i. 4y and 10y

12. For Exercises **a-e**, △ADE is similar to △ABC and FDCE is a square. Find the unknown lengths.
≪See Example 7-8≫

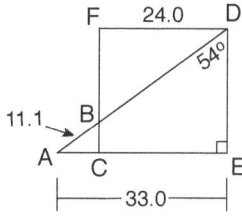

a. Find the length of DE.

b. Find the length of CE.

c. Find the length of AC.

d. Find the length of BC.

e. Find the length of BD.

13. For Exercises **a-f**, △PRQ is similar to △PTS.
≪See Example 7-8≫

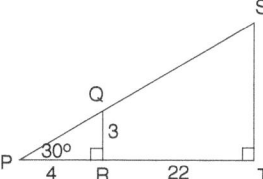

a. m∠PQR   b. m∠PST

c. $\overline{PQ}$ d. $\overline{QS}$

e. $\overline{PS}$ f. $\overline{ST}$

14. Each pair of polygons are similar. Find the values of x and y.
≪See Example 7-9≫

a. 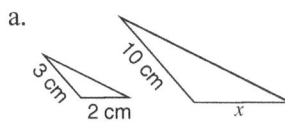   b. 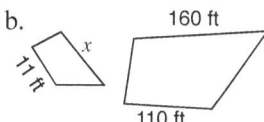   c.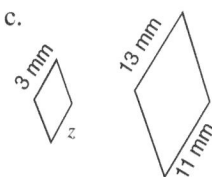

d. 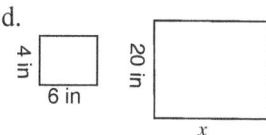   e. 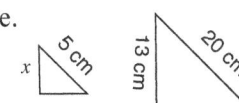   f.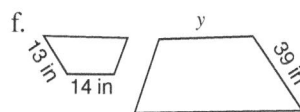

15. In the diagram, given that $\dfrac{EF}{BC} = \dfrac{GF}{CF}$, find the length of CF.
《See Example 7-11》

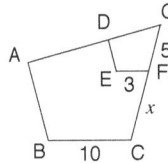

16. For Exercises **a** and **b**, each pair of the triangles are similar, find the value of *x*.
《See Example 7-9》

a. 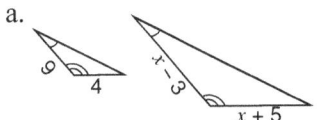   b.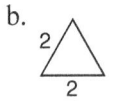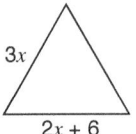

17. For Exercises **a-d**, the triangles below are similar. Find each value.
《See Examples 7-13 and 7-14》

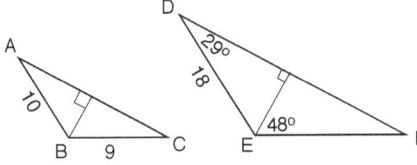

a. Find EF.   b. Find $m\angle A$.

c. Find $m\angle F$.   d. Find $m\angle C$.

18. For Exercises **a-f**, the figures below are similar. Find each value.
《See Examples 7-9 to 7-11》

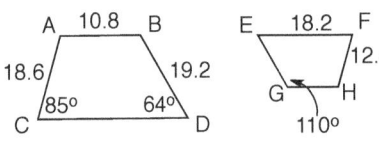

a. $m\angle A$ b. $m\angle B$ c. CD

d. EG   e. GH   f. $m\angle F$

19. In Exercises **a-d**, the polygons are similar to each other. Write a similarity statement that includes the corresponding angles and sides.
≪See Example 7-9≫

a.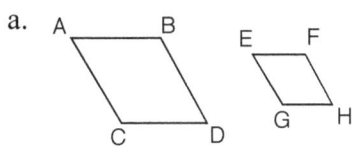

b.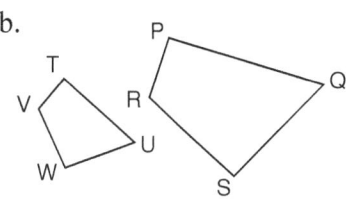

c.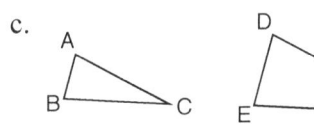

d.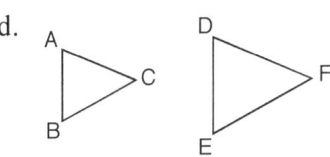

20. In the diagrams, ABCD and GHIJF are similar. Find the values.
≪See Example 7-11≫

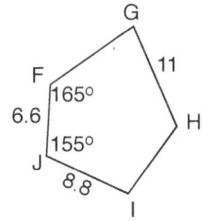 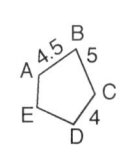

a. $m\angle A$ b. FG

c. DE   d. HI

e. $m\angle E$

21. Tell whether the corresponding angles are congruent. Explain why or why not.
≪See Example 7-11≫

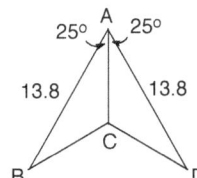

22. Tell whether the corresponding angles are congruent. Explain why or why not.
≪See Example 7-11≫

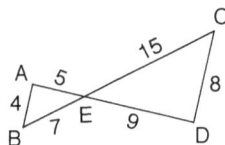

23. △ABC and △DEF are similar.
≪See Examples 7-8 and 7-9≫

    a. $m\angle C$ b. $m\angle D$

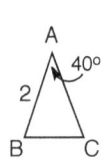 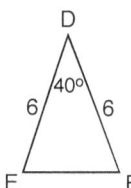   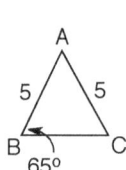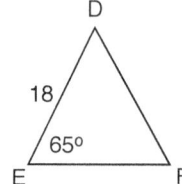

    c. Find the length of DE.   d. $m\angle A$

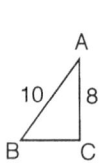 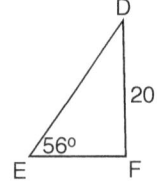   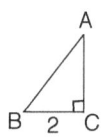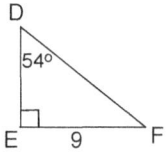

24. Find the length of x, given that each pair of polygons is similar.
≪See Examples 7-8 to 7-11≫

  a. 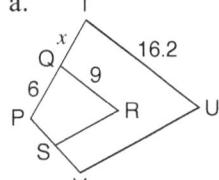   b. 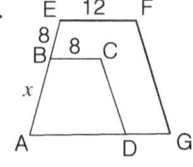   c.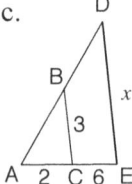

  d. 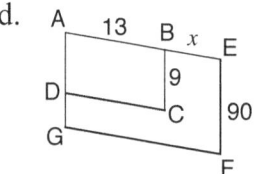   e. 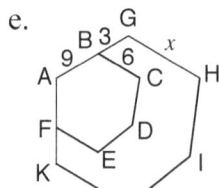   f.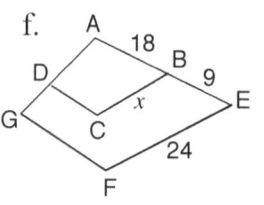

Chapter 7 Similarity

25. Determine whether each pair of figures are similar.
≪See Example 7-9≫

a.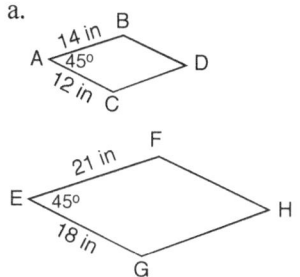

b.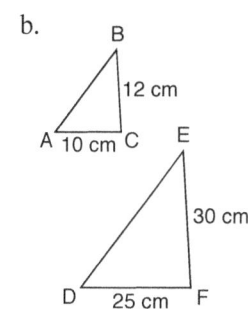

c.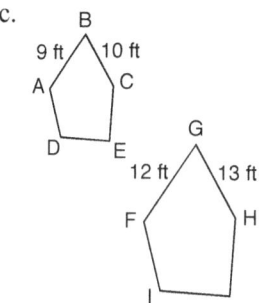

26. For Exercises **a** and **b**, verify whether BC and DE are parallel.
≪See Example 7-8≫

a.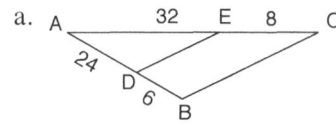

b.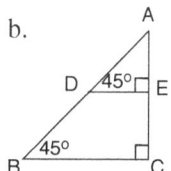

27. For Exercises **a-c**, the triangles shown below are similar. Find each value.
≪See Examples 7-8 and 7-10≫

a. Find the length of BC.

b. Find the length of PR.

c. Find the scale factor of the triangles.

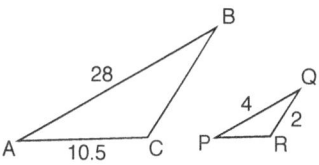

28. For Exercises **a-e**, △ADE is similar to △ABC. Find each value.
≪See Example 7-13≫

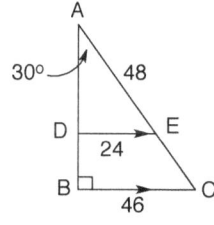

a. Find the length of AD.

b. Find the length of AC.

c. Find the length of BD.

d. Find $m\angle C$.

e. Find $m\angle E$.

29. For Exercises **a-f**, determine whether if each pair of triangles are similar.
≪See Example 7-13≫

a.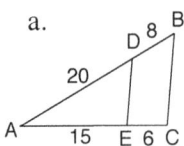
b.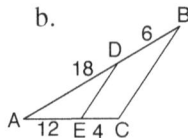
c.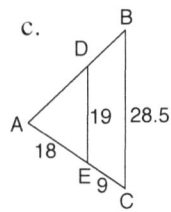

d.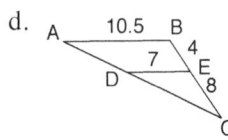
e.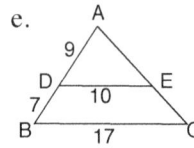
f.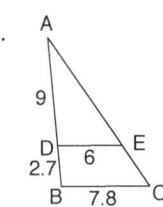

30. Prove that △ABC and △BDC are similar.
≪See Example 7-16≫

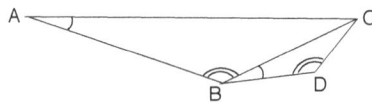

31. For Exercises **a-f**, given that each pair of triangles are similar, find the value of x.
≪See Examples 7-12 to 7-17≫

a.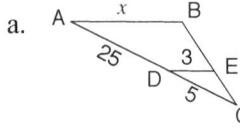
b.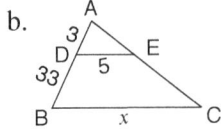
c.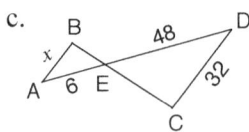

d.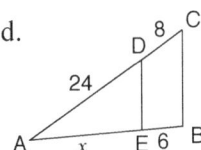
e.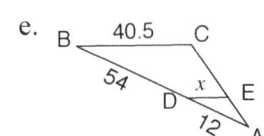
f.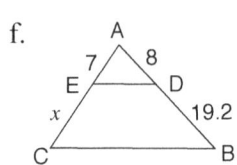

32. For Exercises **a** and **b**, each pair of triangles are similar.
《See Example 7-15》

   a. The perimeter of △ABC is 52.5 cm. Find the perimeter of △DEF.

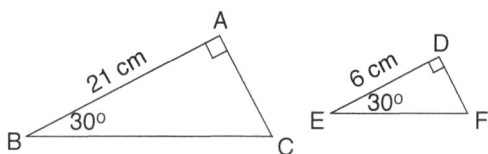

   b. The area of △DEF is 6.4 cm². Find the area of △ABC.

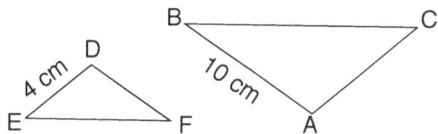

33. For Exercises **a-c**, if each pair of triangles are similar, find the value of x.
《See Examples 7-12 to 7-15》

a. 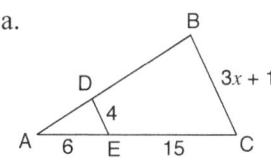   b. 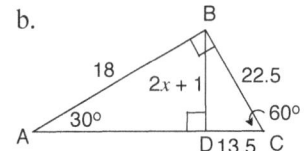   c.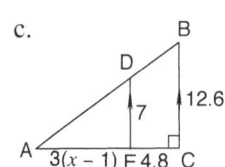

34. Tell whether the figures have corresponding sides that are proportional and corresponding angles that are congruent.
《See Examples 7-14 and 7-15》

a. 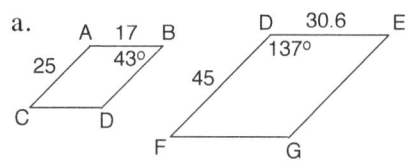   b.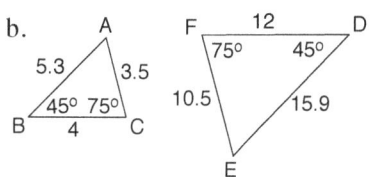

c. 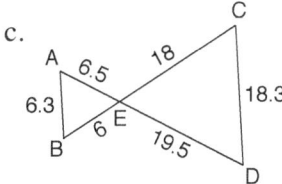   d.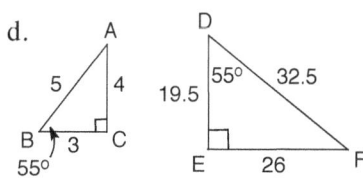

35. For Exercises **a-e**, the triangles are similar.
《See Examples 7-13 to 7-15》

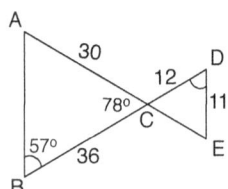

a. Find the length of AB.

b. Find the length of BC.

c. Find $m\angle A$.

d. Find $m\angle D$.

e. Find $m\angle B$.

36. For Exercises **a-i**, if each pair of triangles are similar, BC ∥ DE, find the value of x.
《See Examples 7-12 to 7-17》

a. DE ∥ BC

b. ST ∥ QR

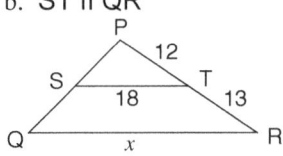

c. BC ∥ DE

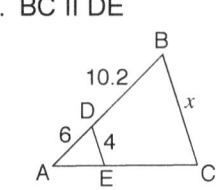

d. ST ∥ PQ

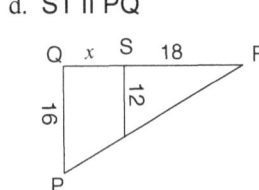

e. VW ∥ TU

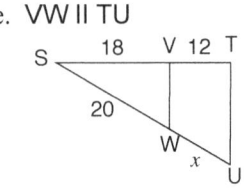

f.

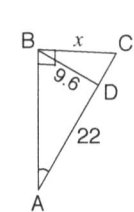

g. CE = 14, BD = 15, and DF = 12

h.

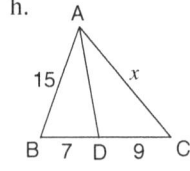

i. AB ∥ CD ∥ EF

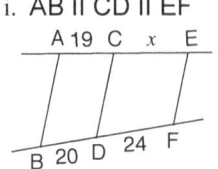

37. For Exercises **a-g**, AB ∥ CD ∥ EF.
≪See Examples 7-18 to 7-20≫

   a. Find the length of EG.

   b. Find ∠3.

   c. Find the length of AC.

   d. Find the length of AB.

   e. Find ∠1.

   f. Find ∠2.

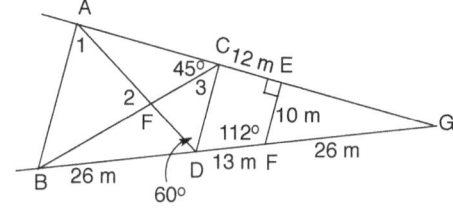

38. For Exercises **a-f**, determine the value of x that makes BC and DE parallel.
≪See Example 7-17≫

   a. 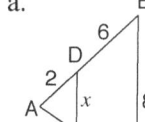   b. 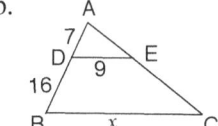   c.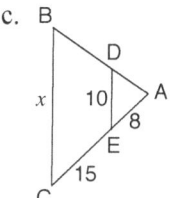

   d. 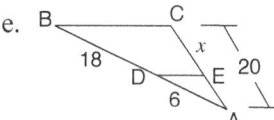   e. 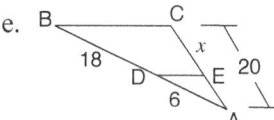   f.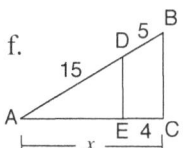

39. Find the scale factor using a dilated image of the figure.
≪See Example 7-21≫

   a. 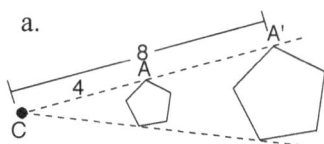   b.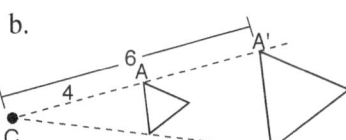

40. Find the dilation of each figure with the origin (0, 0) as the center and the given scale factor.
≪See Example 7-21≫

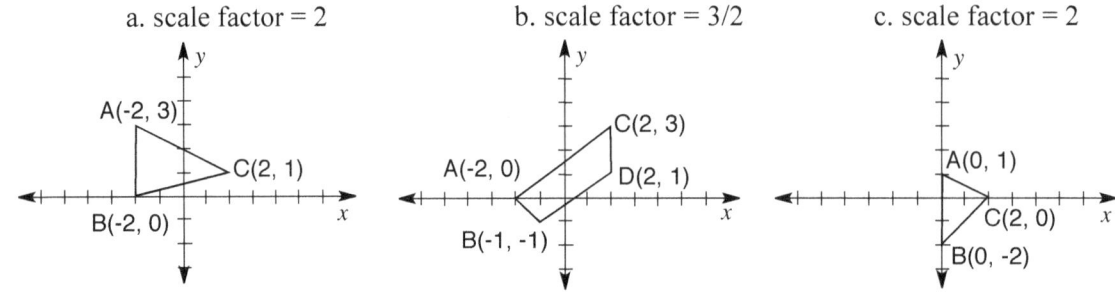

41. Find the dilation of each point with the origin (0, 0) as the center and the given scale factor.
≪See Example 7-21≫

  a. A(4, −1): scale factor of 3

  b. A(−2, −3): scale factor of 4

  c. A(4, 2): scale factor of 3/2

  d. A(0, 0): scale factor of 5

  e. A(2, 0): scale factor of 8

  f. A(−1, 1): scale factor of 2

  g. A(−6, 12): scale factor of 1/3

  h. A(−1, 2): scale factor of 4

42. Find the image of each vertex as a dilation in a coordinate plane with the origin (0, 0) as the center and the given scale factor.
≪See Example 7-21≫

  a. A(2, −1), B(4, −1), C(2, −3), D(4, −3); scale factor 3

  b. P(−3, 3), Q(−1, 2), R(−3, −1), S(−1, 0); scale factor 2

  c. A(−1, 6), B(2, 6), C(2, 0), D(3, 0); scale factor 0.5

  d. P(−4, 3), Q(−1, 2), R(−3, 1), S(−2, 0); scale factor 2

43. For Exercises **a-c**, identify the center of dilation and the scale factor.
《See Example 7-21》

a. 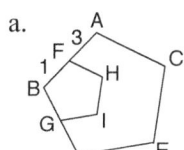   b. 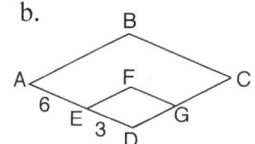   c.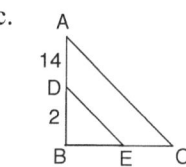

44. Find the image of each vertex as a dilation in a coordinate plane with the origin (0, 0) as the center and the given scale factor.
《See Example 7-21》

a. A(1. −2), B(2, −2), C(1, −3), D(2, , −3); scale factor of 3

b. P(−1, 3), Q(1, 2), R(−1, −1), S(1, 0); scale factor of 3

c. A(−1, 3), B(2, 3), C(2, 0), D(3, 0); scale factor of 2

d. P(−2, 3), Q(−1, 2), R(−3, 1), S(−2, 0); scale factor of 2

45. For Exercises **a-c**, identify the center of the dilation and the scale factor.
《See Example 7-21》

a. 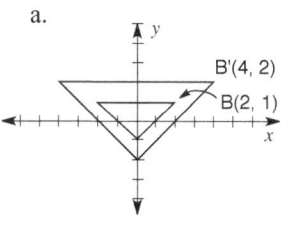   b. 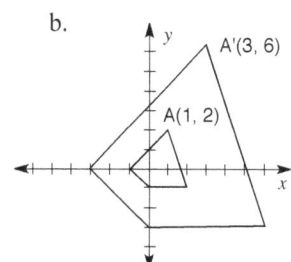   c.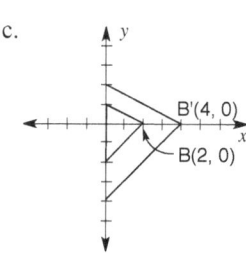

46. Find the image of each point as a dilation with the origin (0, 0) as the center and the given scale factor.
《See Example 7-21》

a. P(1, 5): scale factor 2

b. Q(3, 5): scale factor 3

c. A(8, −2): scale factor 3/2

d. Q(0, 0): scale factor 2

e. P(0, 2): scale factor 4

f. Q(−3, 3): scale factor 5

ANSWERS

(1) a. no, b. yes

(2) a. 5.04, b. 1, c. -4/15, d. -5/4

(3) a. $y = 1.5$, $x = 16$, b. b $x = 32$, $y = 3$

(4) a. slope = -3, b. slope = -3, c. slope = 3/10

(5) a. 1 : 1, b. 2 : 4, 4 : 3, 2 : 3, c. 1 : 6,

(6) a. $\frac{5\,ft}{4\,ft}$, b. $\frac{2\,kg}{1\,kg}$, c. $\frac{10\,cm}{3\,cm}$, d. $\frac{5\,cm}{1\,cm}$, e. $\frac{1\,mi.}{3\,mi.}$, f. $\frac{3\,ft}{1\,ft}$, g. $\frac{1\,oz}{1\,oz}$, h. $\frac{16\,oz}{15\,oz}$, i. $\frac{3\,cm}{50\,cm}$, j. $\frac{2\,cm}{7.5\,cm}$, k. $\frac{4\,in.}{3\,in.}$, l. $\frac{1\,in.}{3\,in.}$

(7) a. 7.2, b. 13.1, c. 0.4, d. 14, e. 4/5, f. 8.7, g. 3.6, h. 7/9, i. -1/2, j. 2.1, k. 5/3, l. 21/2

(8) a. 15, b. 10, c. 16, d. 19.5, e. 78, f. 52.5, g. 5h, h. 20/3

(9) a. 12/4, b. 16/12, c. s/t, d. s/t, e. $(p + 45) / (5 + 45)$, f. $(6 + s) / (t + s)$

(10) a. $x = 12.25$, b. $x = 13/6$, c. $x = 3.6$, d. $x = 10$, e. $x = 18.9$, f. $x = 14/5$

(11) a. 31.6, b. 25.98, c. 22, d. 13.1c, e. 20c, f. 12.8, g. $10x + 50$, h. $20.1y$, i. $6.3y$

(12) a. 24, b. 24, c. 9.0, d. 6.54, e. 29.7

(13) a. 60°, b. 60°, c. 5, d. 27.5, e. 32.5, f. 19.5

(14) a. 20/3, b. 16, c. 33/13, d. 30, e. 3.25, f. 42

(15) 16.7,

(16) a. 11.4, b. 6

(17) a. 16.2, b. 29°, c. 42°, d. 42°

(18) a. 101°, b. 110°, c. 27.3, d. 12.8, e. 7.2, f. 85°

(19) a. $\frac{AB}{EF} = \frac{AC}{EG} = \frac{CD}{GH} = \frac{BD}{FD}$, $\angle A = \angle G$, $\angle B = \angle F$, $\angle C = \angle G$, $\angle D = \angle H$, b. $\frac{TU}{PQ} = \frac{TU}{PR} = \frac{VW}{RS} = \frac{US}{SQ}$, $\angle T = \angle P$, $\angle U = \angle Q$, $\angle V = \angle R$, $\angle W = \angle S$, c. $\frac{AB}{DE} = \frac{BC}{EF} = \frac{AC}{DF}$ $\angle A = \angle D$, $\angle B = \angle E$, $\angle C = \angle F$, d. $\frac{AB}{DE} = \frac{AC}{DF} = \frac{CB}{FE}$ $\angle A = \angle D$, $\angle C = \angle F$, $\angle B = \angle E$,

(20) a. 165°, b. 9.9, c. 4, d. 8.8, e. 155°

(21) a. yes, as both polygons are congruent,

(22) no, as the size lengths are not congruent,

(23) a. 70°, b. 50°, c. 25, d. 54°,

(24) a. $x = 4.8$, b. $x = 16$, c. 12, d. 117, e. 18, f. 16

(25) a. yes, b. yes, c. no

(26) a. yes, b. yes,

(27) a. 14, b. 1.5, c. 7

(28) a. 41.6, b. 92, c. 38.1, d. 60°, e. 60°

(29) a. yes, b. yes, c. yes, d. yes, e. no, f. yes

(30) Triangles ABC and BDC are similar by the AA Congruence Theorem

(31) a. 18, b. 60, c. 4, d. 18, e. 7.4, f. 16.8

(32) a. 15, b. 16

(33) a. 3, b. 8.5, c. 3

(34) a. yes, b. no, c. yes, d. yes

(35) a. 33, b. 10, c. 45°, d. 57°, e. 45°

(36) a. 15.4, b. 37.5, c. 10.8, d. 6, e. 13.3, f. about 10.5, g. 17.5, h. 19.3, i. 22.8

(37) a. GE = 24 b. 45°, c. AC = 24, d. 25, e. 30°, f. 75°

(38) a. 2, b. 29.6, c. 28.75, d. 3.7, e. 10, f. 16

(39) a. 1 : 2, b. 2 : 3

(41) a. A'(12, −3), b. A'(−8, −12), c. A'(−6, 3), d. A'(0, 0), e. A'(16, 0), f. A'(−2, 2), g. A'(−2, 4), h. A'(−4, 8)

(42) a. A'(6, −3), B'(11, −3), C'(6, −4), D'(12, −9), b. P'(−6, 6), Q'(−2, 4), R'(−6, −2), S'(−2, 0), c. A'(−0.5, 3), B'(1, 3), C'(1, 0), D'(1.5, 0), d. P'(6, −3), Q'(−8, 6), R'(−6, 2), S'(−4, 0)

(43) a. 3 : 1, b. 2 : 1, c. 7 : 1

(44) a. A'(3, −6), B'(6, −6), C'(3, −9), D'(6, -9), b. P'(−3, 9), Q'(3, 6), R'(−3, −3), S'(3, 0), c. A'(−2, 6), B'(4, 6), C'(4, 0), D'(6, 0), d. P'(−4, 6), Q'(−2, 4), R'(−6, 2), S'(−4, 0)

(45) a. (0, 0); 2, b. (0, 0); 3, c. (0, 0); 2

(46) a. (2, 10), b. (9, 15), c. (12, −3) , d. (0, 0), e. 0, 8), f. (−15, 15).

(40)

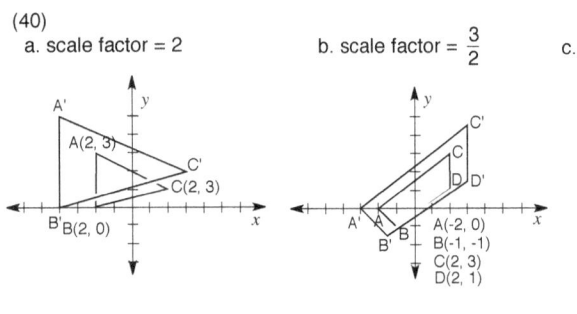
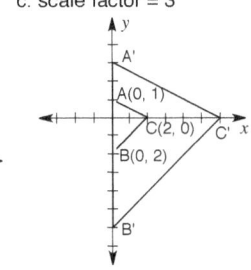

a. scale factor = 2   b. scale factor = $\frac{3}{2}$ c. scale factor = 3

26

SELF-TEST

1. Which of the following can be written as the ratio of a and b?

   (a) $\frac{a}{b}$, where b ≠ 0.
   (b) $a{:}b$
   (c) a to b
   (d) All of the above.

2. Which of the following ratios is equal to 15 : 21?

   (a) 7 to 5
   (b) 12 to 18
   (c) 5 to 7
   (d) 40 to 50

3. Which of the following ratios is not equal to 11 to 48 as x to 22?

   (a) $\frac{11}{x} = \frac{48}{22}$
   (b) $\frac{11}{48} = \frac{x}{22}$
   (c) $\frac{48}{x} = \frac{11}{22}$
   (d) $\frac{11+48}{48} = \frac{x+22}{22}$

4. Which of the following is a proportion?

   (a) $\frac{1}{5}$? $\frac{3}{15}$
   (b) $\frac{2}{3}$? $\frac{14}{21}$
   (c) $\frac{1}{2}$? $\frac{3}{6}$
   (d) All of the above.

5. Which of the following is a proportion?

   (a) $\frac{16}{7} = \frac{7}{16}$
   (b) $\frac{13}{26} = \frac{39}{78}$
   (c) $\frac{2}{5} = \frac{8}{21}$
   (d) $\frac{3}{5} = \frac{6}{5}$

6. The ratios of the lengths in △ABC are 4 : 7 : 5 and the length of the shortest side AC is 19 in. Find the lengths of BC given that the perimeter of △ABC is 76 in.² Round your answer to the nearest tenth.

   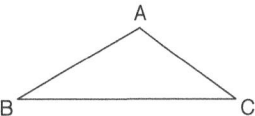

   (a) 33.3 in.
   (b) 23.8 in.
   (c) 42.8 in.
   (d) 38.0 in.

7. The ratios of the angles in △ABC are 9 : 23 : 12. Find the biggest angle of △ABC. Round your answer to the nearest tenth.

   (a) 102.2°
   (b) 94.1°
   (c) 49.0°
   (d) 36.8°

8. If $\frac{r}{6} = \frac{p}{5}$, then $\frac{r}{p} = Q$, which of the following ratios represents the value of Q?

   (a) $\frac{6}{5}$
   (b) $\frac{p+5}{5}$
   (c) $\frac{5}{6}$
   (d) All of the above

27

9. If $\frac{r}{17} = \frac{p}{15}$, then $\frac{r+17}{17} = Q$, which of the following ratios represents the value of Q?

(a) $\frac{r}{p}$
(b) $\frac{17}{15}$
(c) $\frac{p+15}{15}$
(d) All of the above

10. How do you find the value of x?

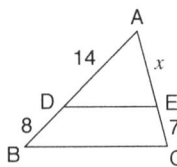

(a) $\frac{14+x}{8} = \frac{8+7}{x}$
(b) $\frac{14+7}{8} = \frac{x+8}{x}$
(c) $\frac{14+8}{14} = \frac{x+7}{x}$
(d) $\frac{14+x}{7} = \frac{14+x}{8}$

11. How do you solve the value of x?

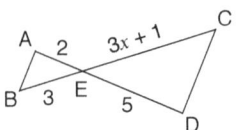

(a) $(2)(3) = (3x+1)(5)$
(b) $\frac{3x+1}{2} = \frac{5}{2}$
(c) $\frac{2}{3} = \frac{3x+1}{5}$
(d) $\frac{3x+1}{3} = \frac{5}{2}$

12. Which of the following has the smallest price per item?

(a) $4 for 2 pounds of apple
(b) $6.50 for 8-pack AA battery
(c) $21 for 2 movie tickets
(d) $12.50 for 3 gallons of gas

13. Which of the following has the highest price per item?

(a) $18.50 for 3 pounds of steak
(b) $16 for 5-pack ice cream
(c) $6.80 for 24 bottles of water
(d) $4.90 for 18 eggs

14. Find the geometric mean of $5c$ and $20c$.

(a) $8c$
(b) $\frac{1}{8c}$
(c) $10c$
(d) $\frac{1}{10c}$

15. Find the geometric mean of $3x$ and $27x$.

(a) $7x$
(b) $8x$
(c) $9x$
(d) $10x$

16. If the lengths of $\triangle ABC$ have 4 cm, 6 cm, and 8 cm and the ratios of the side lengths of $\triangle ABC$ to $\triangle DEF$ are 2 : 7, which of the following is the perimeter of $\triangle DEF$?

(a) 36 cm
(b) 18 cm
(c) 126 cm
(d) 63 cm

17. △ABC is similar to △DEF and the ratio of the perimeter of △ABC to △DEF is 7 : 8. Find the perimeter of △DEF if the perimeter of △ABC is 59.8 in. Round your answer to the nearest tenth.

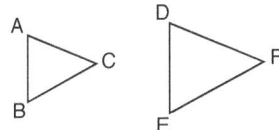

(a) 418.6 in.
(b) 68.3 in.
(c) 52.3 in.
(d) 478.4 in.

18. △ABC is similar to △DEF. The ratios of the perimeter of △ABC to △DEF are 13:7. Find the perimeter of △DEF if the perimeter of △ABC is 140 in. Round your answer to the nearest tenth.

(a) 260.0 in.
(b) 980 in.
(c) 75.4 in.
(d) 1820 in.

19. ABCD is similar to rectangle EFGH. Find the length of CD.

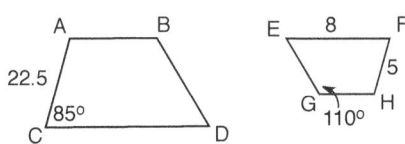

(a) 24
(b) 28
(c) 32
(d) 36

20. △ABC is similar to △DEF. Find the length of DE.

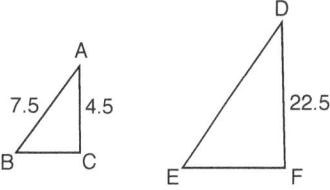

(a) 28.5
(b) 32.5
(c) 37.5
(d) 39.5

21. Which of the following statements is a definition of two similar polygons?

(a) If two polygons are similar with a scale factor of $\frac{5}{6}$, then the perimeter of the polygons have a ratio of $\frac{5}{6}$.
(b) If two polygons are similar, then their corresponding angles are congruent.
(c) If two polygons are similar, then the measures of their corresponding sides are proportional.
(d) All of the above.

22. Which of the following statements is a definition for any two polygons?

(a) If the corresponding angles of two polygons are congruent, then the ratios of the lengths of the corresponding sides are equal.
(b) If one pair of corresponding side of two polygons is equal, then their polygons are similar.
(c) If the ratios of two polygons are similar, then their polygons are similar.
(d) All of the above.

23. If △ABC ~ △DEF with a scale factor of $\frac{3}{10}$, which of the following is the ratio of the perimeter of △ABC to △DEF?

(a) $\frac{10}{3}$
(b) $\frac{3}{10}$
(c) $\frac{6}{10}$
(d) None of the above

24. If △ABC ~ △DEF with a scale factor of $\frac{6}{7}$ and ∠A corresponds to ∠D, which of the following could be the measure of ∠A if the measure of ∠D is 48°?

(a) 56°
(b) 41°
(c) 48°
(d) All of the above.

25. If △ABC ~ △DEF with a scale factor ABC to DEF is $\frac{5}{3}$, the length of AB corresponds to DE, which of the following could be the length of DE if the length of AB is 54 ft?

(a) 90 ft
(b) 32.4 ft
(c) 54 ft
(d) All of the above.

26. Which of the following is a correct definition for any two triangles?

(a) If the scale factors of △ABC and △DEF are equal but the sizes of △ABC and △DEF are different, then △ABC and △DEF are similar.
(b) If two angles of △ABC are congruent to two angles of △DEF, then △ABC and △DEF are similar.
(c) If the ratios of the lengths △ABC and △DEF are equal to the scale factor, then △ABC and △DEF are similar.
(d) All of the above.

27. Given that △ABE is similar to △DCE, find the length of EC.

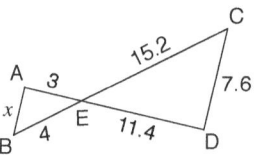

(a) 1.8
(b) 2.0
(c) 2.2
(d) 2.4

28. If BC ∥ DE, find the length of AE.

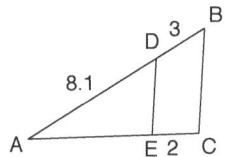

(a) 7.5
(b) 6.7
(c) 5.9
(d) 5.4

29. In the diagram, which of the following is a true statement for the two similar triangles below?

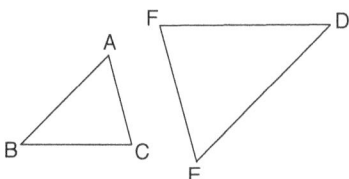

(a) If the lengths of the corresponding sides of △ABC and △DEF are proportional, then △ABC and △DEF are similar.
(b) If $\frac{AB}{DE} = \frac{BC}{EF} = \frac{AC}{DF}$, then △ABC ~ △DEF.
(c) If ∠A ≅ ∠D and $\frac{AB}{DE} = \frac{AC}{DF}$, then △ABC ~ △DEF
(d) All of the above.

30. If △ABE ~ △DCE, find the length of AE.

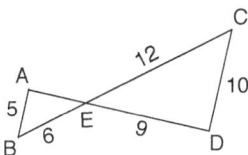

(a) 3
(b) 4
(c) 4.5
(d) 5.5

31. △ABC is similar to △DEF. Find the length of DF.

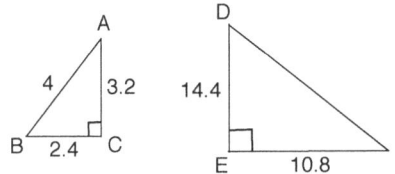

(a) 11
(b) 15
(c) 18
(d) 20

32. If $m\angle R = m\angle T = 80°$, find the length of UT.

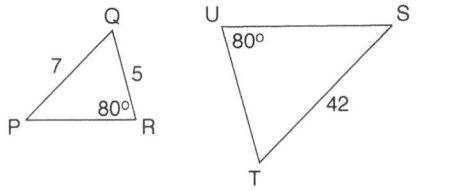

(a) 40
(b) 35
(c) 30
(d) 20

33. If $m\angle R = m\angle T = 78°$, BC ≅ DC, AC ≅ EC, which of the following is the length of DE?

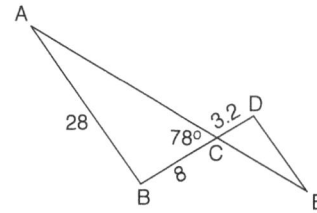

(a) 11.2
(b) 12.2
(c) 13.2
(d) 14.2

34. If △ABC ~ △DEF and $\frac{AB}{DE} = \frac{5}{7}$, which of the following is the length of AC given that DF = 8.4?

(a) 3
(b) 4
(c) 5
(d) 6

35. Which of the following is not necessary in order to prove that △ABC ~ △ADE?

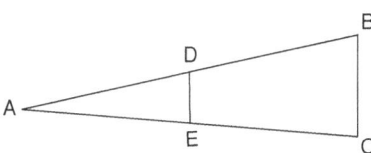

(a) ∠A ≅ ∠A
(b) BC ∥ DE
(c) $\frac{AD}{AB} = \frac{AE}{AC}$
(d) None of the above

36. Which of the following is necessary in order to show that △ABC ~ △DEF?

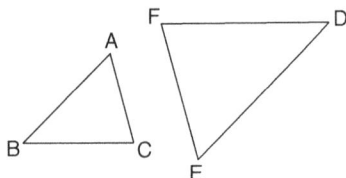

(a) All of the corresponding angles are equal.
(b) The ratios of the side lengths of △ABC and △DEF are equal.
(c) △ABC and △DEF are not congruent.
(d) None of the above

37. What is the length of BD if △ABC ~ △ADE?

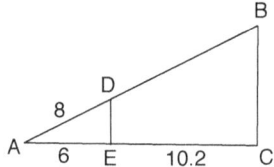

(a) 13.2
(b) 13.6
(c) 14.2
(d) 14.6

38. In the diagram, which of the following statements about the two triangles below are true?

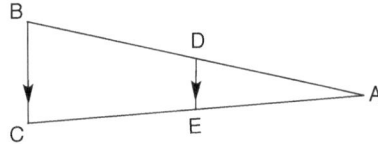

(a) If a line is parallel to one side of △ABC and intersects the other two sides, then it divides the two sides proportionally.
(b) If $\frac{AD}{BD} = \frac{AE}{CE}$, then BC ∥ DE.
(c) If BC ∥ DE, then $\frac{AD}{BD} = \frac{AE}{CE}$.
(d) All of the above.

39. If ∠A ≅ ∠A and BC ∥ DE, find the value of x.

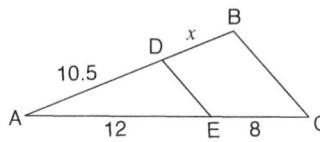

(a) 10.5
(b) 9.5
(c) 8.5.
(d) 7.5

40. If ∠BAD ≅ ∠CAD, find the value of x.

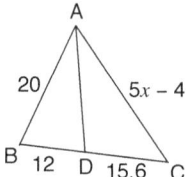

(a) 5
(b) 6
(c) 7
(d) 8

41. If $\frac{BD}{DF} = \frac{7}{3}$, what is the length of CE that makes AC and CE proportional to BD and DF?

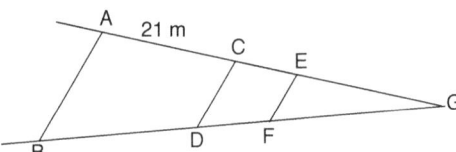

(a) 5
(b) 6
(c) 7
(d) 8

42. △ABC has vertices A(−2, 2), B(3, 2), and C(1, −2). Which of the following are the coordinates after the triangle is dilated given the scale factor is 2?

(a) A'(−4, 4)
(b) B'(6, 4)
(c) C'(2, −4)
(d) All of the above

43. Which of the following is the scale factor given the polygon is being dilated?

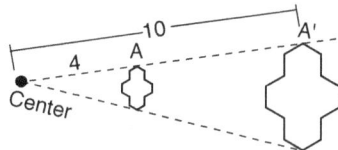

(a) $\frac{2}{5}$
(b) $\frac{5}{2}$
(c) $\frac{3}{2}$
(d) $\frac{2}{3}$

44. Which of the following is the scale factor of the two shapes below?

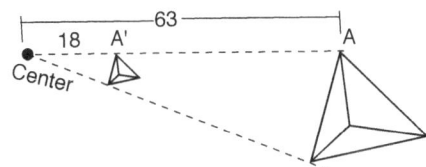

(a) $\frac{2}{5}$
(b) $\frac{5}{2}$
(c) $\frac{2}{7}$
(d) $\frac{7}{2}$

45. Which of the following is the center and scale factor of the two shapes below?

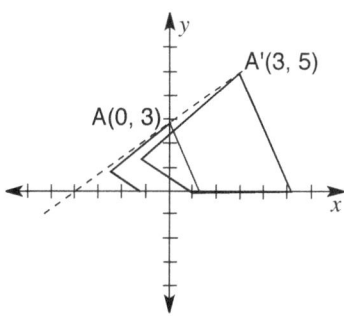

(a) c(−4, 0) and scale factor = $\frac{3.6}{5}$
(b) c(−4, 0) and scale factor = $\frac{5}{3.6}$
(c) c(−4, 0) and scale factor = $\frac{5}{8.6}$
(d) c(−4, 0) and scale factor = $\frac{8.6}{5}$

46. Which of the following is the center and scale factor of the two shapes below?

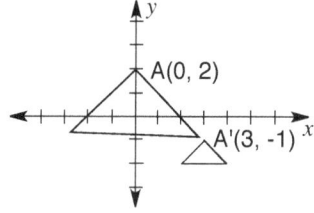

(a) c(4, −2) and scale factor = $\frac{2}{5}$
(b) c(5, −3) and scale factor = $\frac{5}{2}$
(c) c(5, −3) and scale factor = $\frac{2}{5}$
(d) c(4, −2) and scale factor = $\frac{5}{2}$

47. Which of the following is the center and scale factor of the two shapes below?

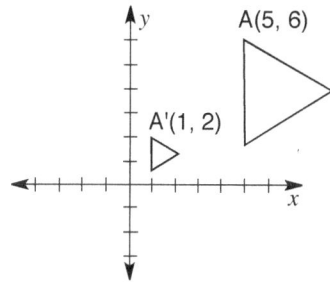

(a) c(−1, 0) and scale factor = 3
(b) c(−1, 0) and scale factor = $\frac{2}{3}$
(c) c(0, 0) and scale factor = $\frac{3}{2}$
(d) c(0, 0) and scale factor = 3

ANSWERS

(1) d	(2) c	(3) c	(4) d	(5) b	(6) a
(7) b	(8) a	(9) c	(10) c	(11) d	(12) b
(13) a	(14) c	(15) c	(16) d	(17) b	(18) c
(19) d	(20) c	(21) d	(22) a	(23) b	(24) c
(25) b	(26) d	(27) b	(28) d	(29) d	(30) c
(31) c	(32) c	(33) a	(34) d	(35) d	(36) b
(37) b	(38) d	(39) a	(40) b	(41) c	(42) d
(43) b	(44) c	(45) d	(46) c	(47) a	

CHAPTER 8
Right Triangles

In this chapter, you will learn how to solve problems involving similar triangles formed by the altitude that is drawn to the hypotenuse, find the lengths of the sides of a triangle with the Pythagorean Theorem, use side lengths to classify triangles and angles measures and find their length in special right triangles, and find the magnitude and direction of a vector. Most importantly, you will learn about trigonometric ratios and how to apply them into finding the side lengths of a triangle.

CONCEPTS   EXAMPLES   FORMULAS   VOCABULARY

8-1. $\triangle ABD$ and $\triangle CBD$ are similar right triangles and form $\triangle ABC$.

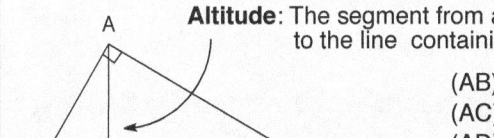

Altitude: The segment from a vertex of a triangle perpendicular to the line containing the opposite side.

$(AB)^2 = (BD)(BC)$
$(AC)^2 = (BC)(CD)$
$(AD)^2 = (BD)(CD)$

The three right triangles are similar by the AA Similarity Postulate so that $\triangle ABD \sim \triangle ACD$, $\triangle ACD \sim \triangle ABC$, and $\triangle ABD \sim \triangle ABC$. AB is the geometric mean of BD and BC, AC is the geometric mean of BC and CD, and AD is the geometric mean of BD and CD. Therefore, you can compare each similar triangle proportionally through their corresponding lengths such as $\frac{BD}{AB} = \frac{AB}{BC}$, $\frac{BC}{AC} = \frac{AC}{CD}$, and $\frac{BD}{AD} = \frac{AD}{CD}$.

8-2. Use the given information in the diagram to find the values of x and y.

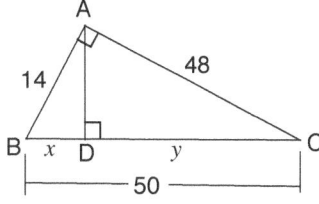

SOLUTION

First, you can draw the three similar right triangles from above and label those triangles as small, medium, or large.

36   Chapter 8   Right Triangles

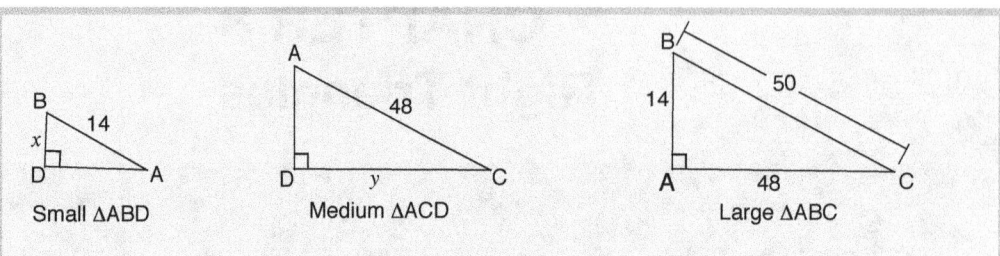

Second, look at the given information in each triangle. △ABD has a shorter leg (*x*) and a hypotenuse of 14. △ADC has a longer leg (*y*) and a hypotenuse of 48. △ABC has a short leg (14), a long leg of 48, and a hypotenuse of 50. △ABC has all the sides mentioned. So it is easier to compare △ABC to △ABD and △ABC to △ACD.

Let's give those sides the notations of Shorter Leg (SL), Longer Leg (LL), and Hypotenuse (Hyp).

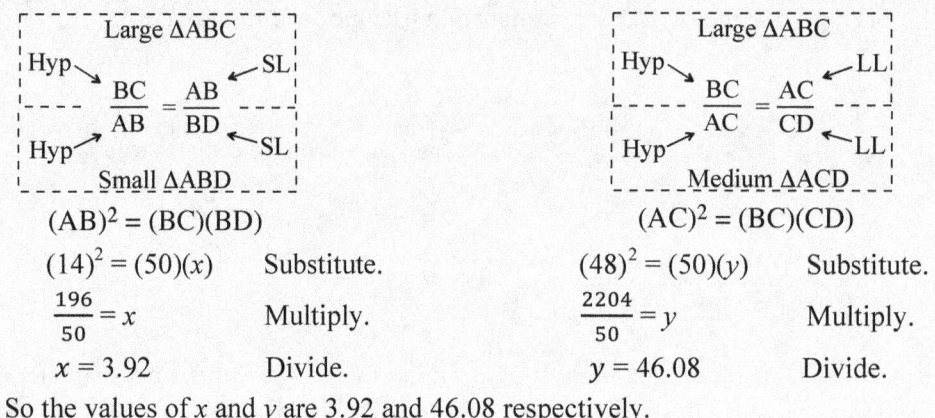

So the values of x and y are 3.92 and 46.08 respectively.

8-3. Use the given information in the diagram to find the value of x.

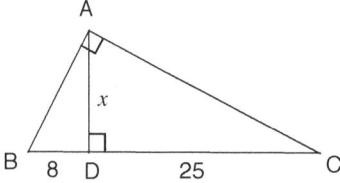

SOLUTION

By the Geometric Mean Theorem, in a right triangle, the altitude from the right angle to the hypotenuse divides the hypotenuse in two segments. The length (AD) of the altitude is the geometric mean of the lengths of the two segments (BD and CD). Therefore, it is easier to compare △ABD to △ADC.

Chapter 8 Right Triangles

```
        Medium △ADC
  LL ↘         ↙ SL
       CD   AD
       ── = ──
       AD   BD
  LL ↗         ↖ SL
        Small △ABD
```

$(AD)^2 = (BD)(CD)$

$(x)^2 = (8)(25)$ Substitute.

$x = \sqrt{(8)(25)}$ Square root.

$x = 10\sqrt{2}$ Simplify.

So the value of x is $10\sqrt{2}$.

8-4. Use the given information in the diagram to find the value of x.

a.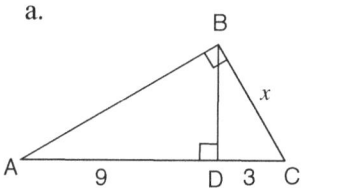

b.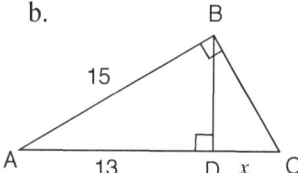

SOLUTION

By the Geometric Mean Theorem, in a right triangle, the altitude from the right angle to the hypotenuse (BC) divides the hypotenuse in two segments (AC and CD). The length of each leg of the right triangle is the geometric mean to the length of the hypotenuse and the segment (AB) of the hypotenuse is adjacent to the leg (AD and AC).

a.
```
        Large △ABC
  Hyp ↘         ↙ SL
       AC   BC
       ── = ──
       BC   CD
  Hyp ↗         ↖ SL
        Small △BCD
```

b.
```
        Large △ABC
  Hyp ↘         ↙ LL
       AC   AB
       ── = ──
       AB   AD
  Hyp ↗         ↖ LL
        Medium △ABD
```

$(BC)^2 = (AC)(CD)$

$(x)^2 = (9+3)(3)$ Substitute.
$x^2 = 36$ Simplify.
$x = 6$ Square root.

$(AB)^2 = (AC)(AD)$

$(15)^2 = (13)(13+x)$ Substitute.
$225 = 169 + 13x$ Simplify.
$225 - 169 = 13x$ Subtract 169 from both sides.
$4.3 = x$ Simplify/Divide.

8-5. Use the given information in the diagram to find the value of x and the area of the triangle.

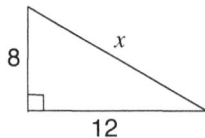

SOLUTION

The Pythagorean Theorem describes the relationship between the length of the hypotenuse of a right triangle to the lengths of the other two legs.

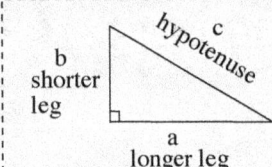

The longest side is called the hypotenuse(c), which is the side opposite the right angle that is adjacent to the shorter and longer legs. The formula of the Pythagorean Theorem is $c^2 = a^2 + b^2$

$c^2 = a^2 + b^2$ Pythagorean Theorem
$x^2 = 12^2 + 8^2$ Substitute.
$x^2 = 144 + 64$ Multiply.
$x^2 = 208$ Add.
$x = 4\sqrt{13}$ Simplify.

So the value of x is $4\sqrt{13}$.

The area of a triangle can be found using the formula $A = (1/2)bh$, where b is the base of the triangle and h is the height.

So $A = (1/2)bh$ Formula of the area of a triangle
$A = (1/2)(12)(8)$ Substitute.
$A = 48$ units^2 Simplify.

8-6. Use the given information in the diagram to find the value of x.

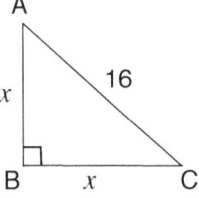

SOLUTION

You can apply the Pythagorean Theorem.
$c^2 = a^2 + b^2$ Pythagorean Theorem
$16^2 = x^2 + x^2$ Substitute.
$256 = 2x^2$ Multiply.
$128 = x^2$ Add.
$8\sqrt{2} = x$ Simplify.

So the value of x is $8\sqrt{2}$.

8-7. Use the given information in the diagram to find the value of x.

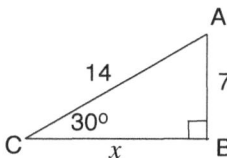

SOLUTION

Use the Pythagorean Theorem to solve the value of x.
$c^2 = a^2 + b^2$ Pythagorean Theorem
$14^2 = 7^2 + x^2$ Substitute.
$196 = 49 + x^2$ Multiply.
$147 = x^2$ Add.
$7\sqrt{3} = x$ Simplify.
So the value of x is $7\sqrt{3}$.

8-8. Is the following triangle a right triangle? Explain your answer.

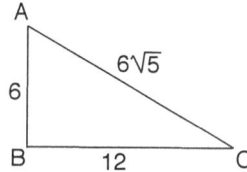

SOLUTION

By the Converse of the Pythagorean Theorem, if the square of the length of the hypotenuse is equal to the sum of the squares of the lengths of the other two legs, then the triangle is a right triangle. If $c^2 = a^2 + b^2$, then $\triangle ABC$ is a right triangle.

c^2 ? $a^2 + b^2$
$(6\sqrt{5})^2$? $6^2 + 12^2$
180 ? $36 + 144$
$180 = 180$
So because $c^2 = a^2 + b^2$, $\triangle ABC$ is a right triangle.

8-9. Classify the triangle as right, acute, or obtuse.

a.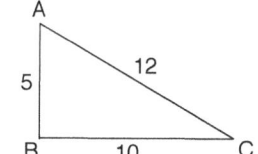

b.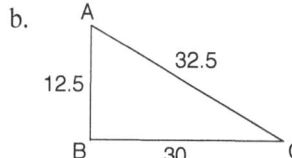

Chapter 8 Right Triangles

SOLUTION

You can compare the square of the hypotenuse with the sum of the squares of the other legs. If $c^2 = a^2 + b^2$, then $\triangle ABC$ is a right triangle, if $c^2 < a^2 + b^2$, then $\triangle ABC$ is acute, and if $c^2 > a^2 + b^2$, then $\triangle ABC$ is obtuse.

a) $\quad c^2 \;?\; a^2 + b^2$	b) $\quad c^2 \;?\; a^2 + b^2$
$\quad 12^2 \;?\; 5^2 + 10^2$	$\quad 32.5^2 \;?\; 12.5^2 + 30^2$
$\quad 144 \;?\; 25 + 100$	$\quad 1056.25 \;?\; 156.25 + 900$
$\quad 144 > 125$	$\quad 1056.25 = 1056.25$
As $c^2 > a^2 + b^2$, $\triangle ABC$ is an obtuse triangle.	As $c^2 = a^2 + b^2$, $\triangle ABC$ is a right triangle.

8-10. Determine whether each triangle is right, acute, or obtuse.

a. $10, 7\sqrt{2}, 6\sqrt{3}$

b. $5, 2, 7$

SOLUTION

You must find the square of the lengths of the hypotenuse and the two legs.

$10^2 = 100$, $(7\sqrt{2})^2 = 98$, $(6\sqrt{3})^2 = 108$. You know that $6\sqrt{3}$ is the length of the hypotenuse because the longest length is called the hypotenuse. Now you can use the equations $c^2 = a^2 + b^2$ (right triangle), $c^2 < a^2 + b^2$ (acute triangle), or $c^2 > a^2 + b^2$ (obtuse triangle) to determine what kind of triangles they are.

a) $\quad c^2 \;?\; a^2 + b^2$	b) $c^2 \;?\; a^2 + b^2$
$(6\sqrt{3})^2 \;?\; 10^2 + (7\sqrt{2})^2$	$\quad 7^2 \;?\; 5^2 + 2^2$
$\quad 108 \;?\; 100 + 98$	$\quad 49 \;?\; 25 + 4$
$\quad 108 < 198$	$\quad 49 > 29$
As $c^2 < a^2 + b^2$, it is an acute triangle.	As $c^2 > a^2 + b^2$. It is an obtuse triangle.

8-11. Find the values of x and y. Write your answer in the simplest form.

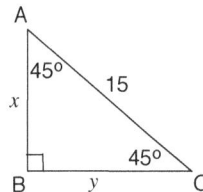

> **SOLUTION**
>
> You can find the unknown lengths of certain triangles by using the Special Right Triangles Theorem.
>
>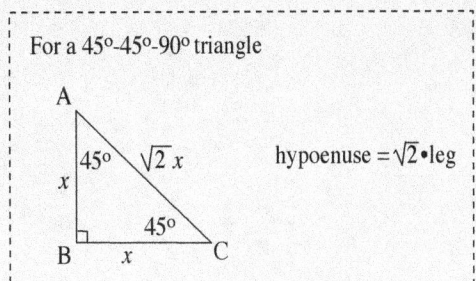
>
> $\triangle ABC$ is a 45°-45°-90° right triangle. So the lengths x and y are equal and the hypotenuse is $\sqrt{2}$ times the length of the shorter leg.
>
> $15 = \sqrt{2}(x)$
> $x = \dfrac{15}{\sqrt{2}}$
> $x = \dfrac{15\sqrt{2}}{2}, \quad y = \dfrac{15\sqrt{2}}{2}$
>
> So the values of x and y are $\dfrac{15\sqrt{2}}{2}$ and $\dfrac{15\sqrt{2}}{2}$ respectively.

8-12. Find the value of x.

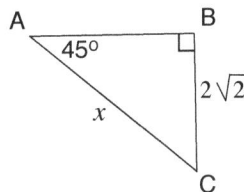

> **SOLUTION**
>
> $\triangle ABC$ is a 45°-45°-90° right triangle. So the length x is the hypotenuse, which is $\sqrt{2}$ times the length of the leg.
>
> Hypotenuse = $\sqrt{2}$ (shorter leg)
> $x = \sqrt{2}\,(2\sqrt{2})$
> $x = 4$
>
> So the value of x is 4.

8-13. Find the values of x and y.

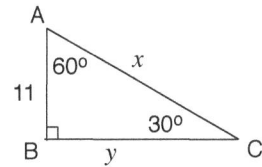

> **SOLUTION**
>
> You can find the unknown lengths of certain triangles by using the Special Right Triangles Theorem.
>
> For a 30°-60°-90° triangle
>
>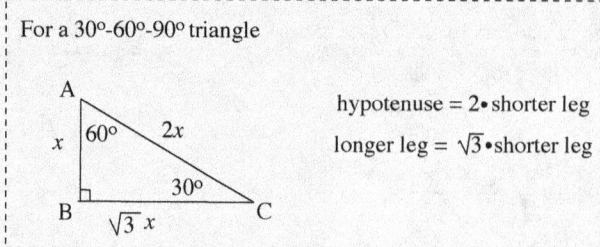
>
> hypotenuse = 2•shorter leg
> longer leg = √3•shorter leg
>
> △ABC is a 30°-60°-90° right triangle. The x represents the length of the shorter leg and the hypotenuse is twice the length of the shorter leg.
>   hypotenuse = 2(shorter leg)
> $x = (2)(11) = 22$
>
> The longer leg is equal to √3 times the length of the shorter leg.
>   Longer leg = √3 shorter leg
> $y = 11\sqrt{3}$
>
> So the values of x and y are 22 and $11\sqrt{3}$.

8-14. **S**oh-**C**ah-**T**oa is a mnemonic used to remember the trigonometry ratio.

Soh-**C**ah-**T**oa is an easy word to remember.

o = opposite, a = adjacent, h = hypotenuse

Soh $\sin A = \dfrac{\text{opposite}}{\text{hypotenuse}}$

Cah $\cos A = \dfrac{\text{adjacent}}{\text{hypotenuse}}$

Toa $\tan A = \dfrac{\text{opposite}}{\text{adjacent}}$

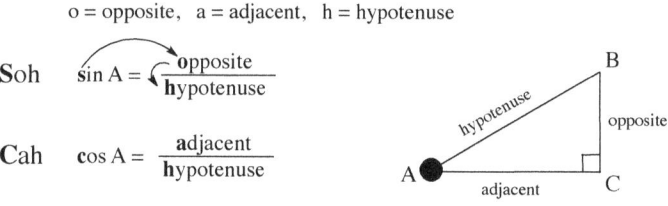

* The longest length is always hypotenuse.

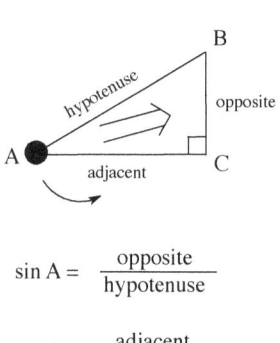

$\sin A = \dfrac{\text{opposite}}{\text{hypotenuse}}$

$\cos A = \dfrac{\text{adjacent}}{\text{hypotenuse}}$

$\tan A = \dfrac{\text{opposite}}{\text{adjacent}}$

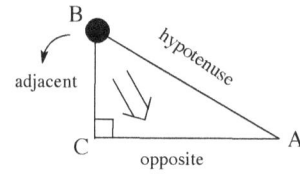

$\sin B = \dfrac{\text{opposite}}{\text{hypotenuse}}$

$\cos B = \dfrac{\text{adjacent}}{\text{hypotenuse}}$

$\tan B = \dfrac{\text{opposite}}{\text{adjacent}}$

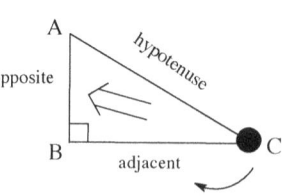

$\sin C = \dfrac{\text{opposite}}{\text{hypotenuse}}$

$\cos C = \dfrac{\text{adjacent}}{\text{hypotenuse}}$

$\tan C = \dfrac{\text{opposite}}{\text{adjacent}}$

8-15. Find the sine, cosine, and tangent of A and C.

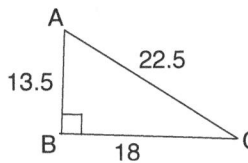

SOLUTION

First, you should find where the hypotenuse is. The hypotenuse is always the longest side of a right triangle and is opposite the right angle. Second, find which leg is adjacent or opposite. In △ABC, the longest leg is AC, which is the hypotenuse. The adjacent leg of ∠A is AB and the opposite leg of ∠A is BC. The leg adjacent to ∠C is BC and leg opposite to ∠C is AB.

$$\sin A = \frac{\text{opposite}}{\text{hypotenuse}} = \frac{18}{22.5} \qquad \sin C = \frac{\text{opposite}}{\text{hypotenuse}} = \frac{13.5}{22.5}$$

$$\cos A = \frac{\text{adjacent}}{\text{hypotenuse}} = \frac{13.5}{22.5} \qquad \cos C = \frac{\text{adjacent}}{\text{hypotenuse}} = \frac{18}{22.5}$$

$$\tan A = \frac{\text{opposite}}{\text{adjacent}} = \frac{18}{13.5} \qquad \tan C = \frac{\text{opposite}}{\text{adjacent}} = \frac{13.5}{18}$$

8-16. Reciprocal Functions.

$$\sin A = \frac{\text{opposite}}{\text{hypotenuse}} \qquad \sin^{-1} A = \csc A = \frac{\text{hypotenuse}}{\text{opposite}}$$

$$\cos A = \frac{\text{adjacent}}{\text{hypotenuse}} \qquad \cos^{-1} A = \sec A = \frac{\text{hypotenuse}}{\text{adjacent}}$$

$$\tan A = \frac{\text{opposite}}{\text{adjacent}} \qquad \tan^{-1} A = \cot A = \frac{\text{adjacent}}{\text{opposite}}$$

8-17. Find the values of *x* and *y*.

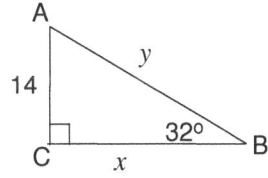

> **SOLUTION**
>
> You should know what measures are given and what measures are missing from the question. The values of AC and ∠B are given while the values of x and y are missing. The y represents the length of the hypotenuse. So you can solve for y by using the sine ratio and find the length x by using the tangent ratio.
>
> $\sin B = \dfrac{\text{opposite}}{\text{hypotenuse}}$ sine ratio $\tan B = \dfrac{\text{opposite}}{\text{adjacent}}$ tangent ratio
> $\sin 32° = \dfrac{14}{y}$ Use the sine ratio. $\tan 32° = \dfrac{14}{x}$ Use the sine ratio.
> $(y)\sin 32° = 14$ Multiply both sides. $(x)\sin 32° = 14$ Multiply both sides.
> $y = \dfrac{14}{\sin 32}$ Divide sin 32° from both sides. $x = \dfrac{14}{\tan 32}$ Divide tan 32° from both sides.
> $y \cong 26.42$ $x \cong 22.4$
>
> So the values of y and x are about 26.42 and 22.2 respectively.

8-18. Find the values of the variables.

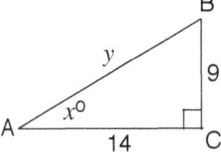

> **SOLUTION**
>
> | Find the measure of x by using the tangent ratio. | Find the value of y by using the tangent ratio. |
> |---|---|
> | $\tan A = \dfrac{\text{opposite}}{\text{adjacent}}$ tangent ratio | $\sin A = \dfrac{\text{opposite}}{\text{hypotenuse}}$ sine ratio |
> | $\tan A = \dfrac{9}{14}$ Use the tangent ratio. | $\sin 33° = \dfrac{9}{y}$ Use the sine ratio. |
> | $\tan A \approx 33°$ Solve angle x. | $(y)\sin 33° = 9$ Multiply both sides. |
> | So the angle x is 33° | $y = \dfrac{9}{\sin 33}$ Divide both sides by sin 33°. |
> | | $y \approx 16.53$ So the value of y is about 16.53. |

8-19. Find the direction of the vector.

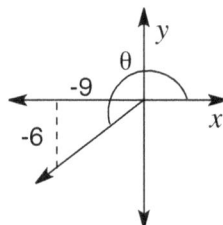

SOLUTION

Find the distance by using the Distance Formula. $d = \sqrt{x^2 + y^2}$ $d = \sqrt{(-9)^2 + (-6)^2}$ $d = 3\sqrt{13}$	Find the direction by using the tangent ratio. $\tan \theta = \dfrac{\text{opposite}}{\text{adjacent}}$ $\tan \theta = \dfrac{-6}{-9}$ $\theta = 214°$ So the direction in the third quadrant of the coordinate (x, y) is $214°$.

8-20. Find the magnitude and direction of the vector.
$\overrightarrow{AB}$: A(−1, 1), B(−4, 6)

SOLUTION

A vector is a quantity that has magnitude and direction such as velocity, momentum, and force. You can find the magnitude vector by using the Distance Formula.

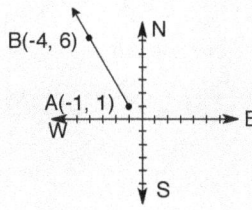

$d = \sqrt{(x_2 - x_1)^2 + (y_2 - y_1)^2}$
$d = \sqrt{((-4) - (-1))^2 + (6 - 1)^2}$
$d = \sqrt{(-3)^2 + 5^2}$
$d = \sqrt{34}$

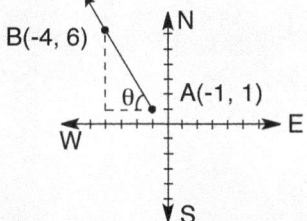

Let (x_1, y_1) be A(−1, 1) and (x_2, y_2) be B(−4, 6).
$\tan \theta = \dfrac{\text{opposite}}{\text{adjacen}}$ Tangent ratio
$\tan \theta = \dfrac{5}{3}$ Use the tangent ratio.
$\theta = 59°$ Solve θ.
So the magnitude of the vector is $\sqrt{34}$ in the direction 59° west of north.

PRACTICES

1. For Exercises **a-d**, answer each question about the triangle below.
《See Examples 8-1 and 8-2》

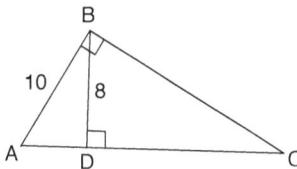

a. Find the length of AD.　b. Find the length of BC.

c. Find the length of AC.　d. Find the length of DC.

2. Use the diagram in the given information to find the value of x in each figure.
《See Examples 8-1 and 8-2》

a. 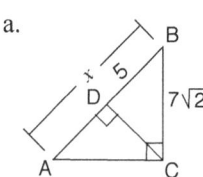　　b. 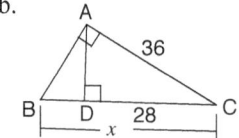　　c.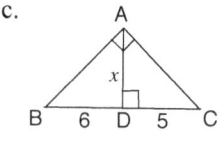

d. 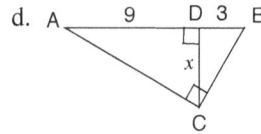　　e. 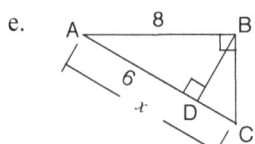　　f.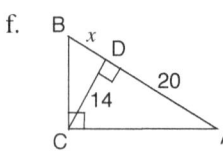

g. 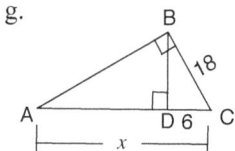　　h. 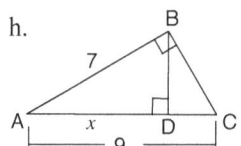　　i.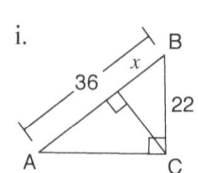

3. Use the given information in the diagram to find each length.
≪See Examples 8-1 to 8-4≫

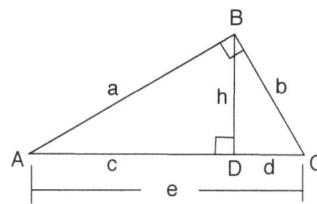

a. If $a = 10$, $b = 6$, and $e = 12$, find the length of h.

b. If $c = 9$ and $d = 4$, find the length of h.

c. If $c = 15$ and $d = 3$, find the length of b.

d. If $c = 12$ and $h = 3$, find the length of e.

e. If $c = 14$ and $e = 18$, find the length of a.

f. If $c = 18$ and $h = 15$, find the length of d.

4. Use the diagrams to find the value of each variable.
≪See Examples 8-1 to 8-4≫

a.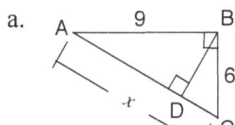

b.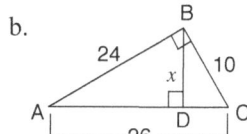

c.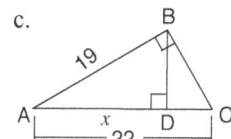

d.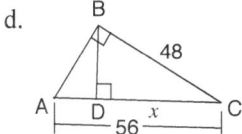

e.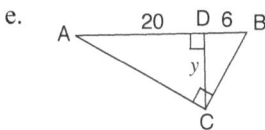

f.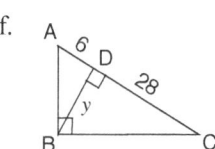

g.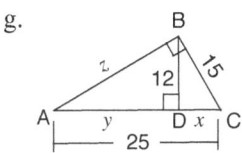

h.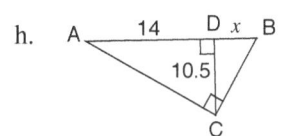

i.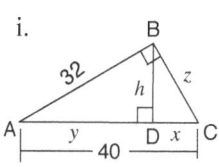

5. Use the given information in the diagram to find each length.
《See Examples 8-1 to 8-4》

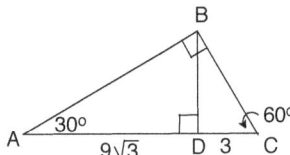

a. Find the length of BD.

b. Find the length of AB.

c. Find the length of BC.

6. Use the diagrams of the right triangles to find the length of each hypotenuse.
《See Example 8-5》

a.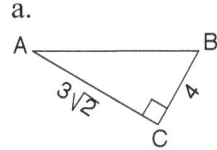

b.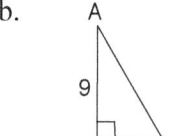

c.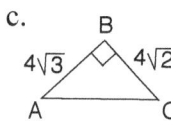

d.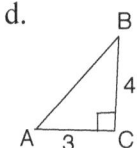

e.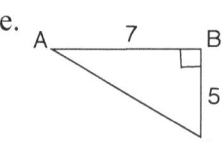

f.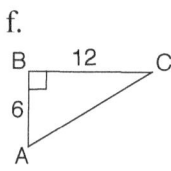

g.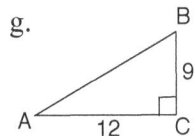

h.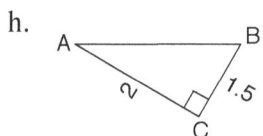

i.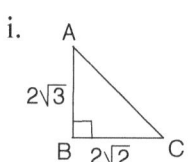

7. Find the area of each right triangle.
《See Example 8-5》

a.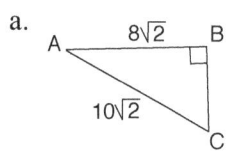

b.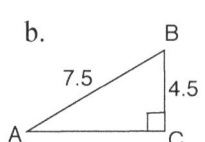

c.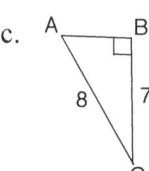

8. For Exercises **a-c**, use the diagram of the square below.
   ≪See Examples 8-5 and 8-6≫

   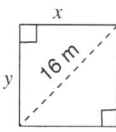

   a. Find the values of *x* and *y*.

   b. Find the area of a triangle.

   c. Find the perimeter of the square.

9. For Exercises **a-c**, use the diagram of the rectangle.
   ≪See Examples 8-5 and 8-6≫

   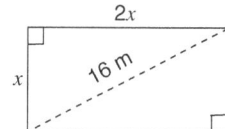

   a. Find the value of *x*.

   b. Find the area of the triangle.

   c. Find the perimeter of the rectangle.

10. Use the given information in the diagram to find each answer.
    ≪See Examples 8-5 to 8-7≫

    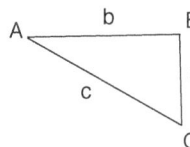

    a. Given that the triangle is right with leg $a = 6$ ft and hypotenuse $c = 20$ ft, find the length of the leg *b*.

    b. If $a = 12$ cm, $b = 16$ cm, find the length of *c* for a right triangle.

    c. If $a = 10$ cm, $b = 12$ cm, and $c = 22$ cm, classify the triangle as right, acute, or obtuse.

    d. If $a = 8$ cm, $b = 10$ cm, and $c = 13$ cm, classify the triangle as right, acute, or obtuse.

    e. If $a = 4.5$ yd, $b = 6$ yd, and $c = 7$ yd, classify the triangle as right, acute, or obtuse.

11. If the length of the side of a square is $6\sqrt{2}$, find the length of the diagonal and the area of the square.
    ≪See Example 8-5≫

    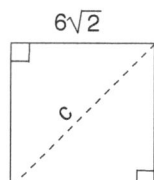

12. Use the diagrams below to classify each triangle as right, acute, or obtuse.
《See Examples 8-8 to 8-10》

a. 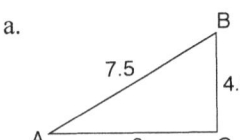   b. 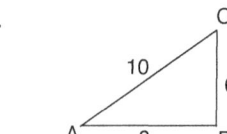   c.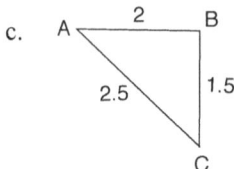

d. 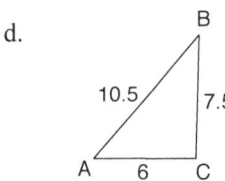   e. 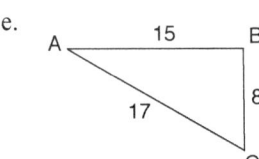   f.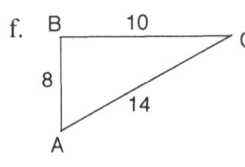

g. 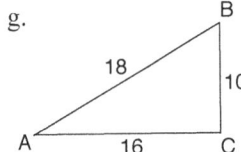   h. 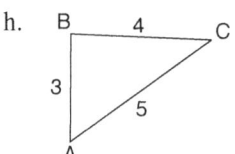   i.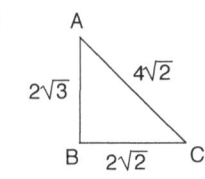

13. For Exercises **a-h**, classify the triangle as right, acute, or obtuse.
《See Example 8-10》

a. $10, 4\sqrt{2}, 8\sqrt{2}$ b. $15, 12, 17$ c. $5, 12, 13$

d. $10, 7\sqrt{2}, 6\sqrt{3}$ e. $2\sqrt{3}, 4\sqrt{3}, 6$ f. $15, 5\sqrt{5}, 3\sqrt{2}$

g. $8\sqrt{2}, 4\sqrt{5}, 12$ h. $10, 24, 26$

14. For Exercises **a-f**, solve the value of each variable.
≪See Example 8-11≫

a.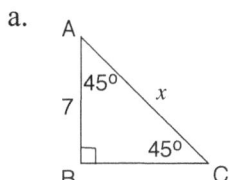

b.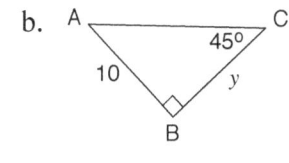

c.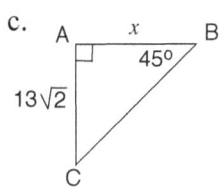

d.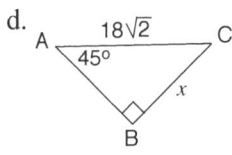

e.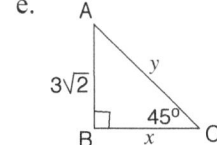

f.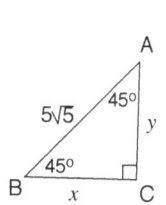

15. If the diameter of a circle is 14 cm, find the length of c and the area of the triangle.
≪See Example 8-5≫

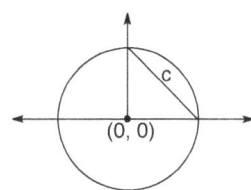

16. For Exercises **a-f**, find the values of x and y.
≪See Example 8-12≫

a.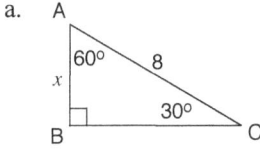

b.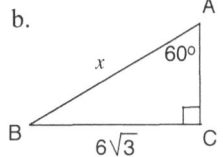

c.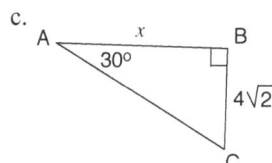

d.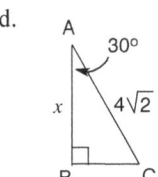

e.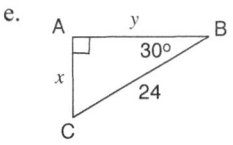

f.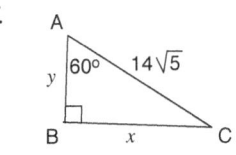

17. For Exercises **a-f**, use the triangle to find each length (scale is not accurate).
《See Example 8-13》

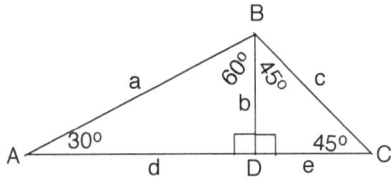

a. If $a = 18$ cm, find the lengths of b and d.

b. If $b = 7$ cm, find the lengths of a and c.

c. If $c = 12$ cm, find the lengths of d and e.

d. If $d = 8\sqrt{3}$ cm, find the lengths of a and b.

e. If $e = 6$ cm, find the lengths of a and c.

f. If $b = 7$cm, find the lengths of d and e.

18. For Exercises **a-f**, find the value of each variable.
《See Examples 8-11 to 8-13》

a.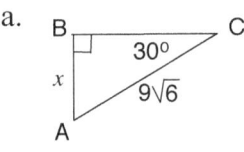

b.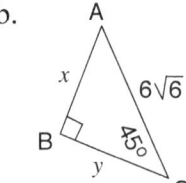

c.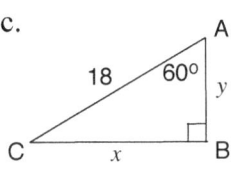

d.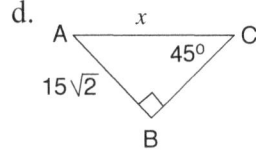

e.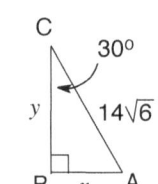

f.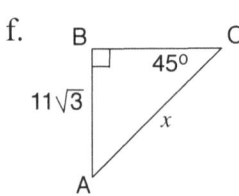

19. Use the given information in the diagram to find the measure of each angle.
《See Example 8-14》

i.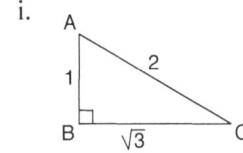

ii.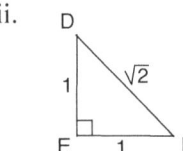

a. $\sin A = \dfrac{\sqrt{3}}{2}$

b. $\cos A = \dfrac{1}{2}$

c. $\tan A = \dfrac{\sqrt{3}}{1}$

d. $\sin C = \dfrac{1}{2}$

e. $\cos C = \dfrac{\sqrt{3}}{2}$

f. $\tan C = \dfrac{1}{\sqrt{3}}$

g. $\sin D = \dfrac{1}{\sqrt{2}}$ h. $\cos D = \dfrac{1}{\sqrt{2}}$ i. $\tan D = \dfrac{1}{1}$

j. $\sin F = \dfrac{1}{\sqrt{2}}$ k. $\cos F = \dfrac{1}{\sqrt{2}}$ l. $\tan F = \dfrac{1}{1}$

20. Use the given information in the diagram to find the solution.
《See Example 8-13》

a. If $a = 12$ cm, find the perimeter of the rectangle.

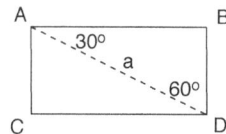

b. If $a = 6\sqrt{3}$ cm, find the perimeter of $\triangle ABC$.

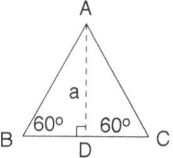

c. If $\overline{AD} = 4$ cm, find the perimeter of $\triangle ABC$.

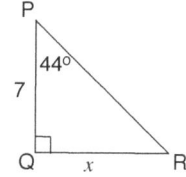

d. If $a = 6\sqrt{2}$ cm, find the perimeter of square ABCD.

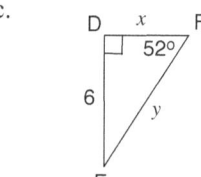

21. For Exercises **a–f**, find the values of each variable. Round your answers to the nearest tenth.
《See Examples 8-14 to 8-17》

a.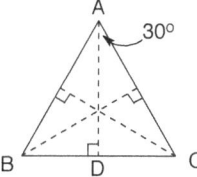

b.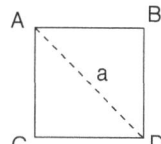

c.

d.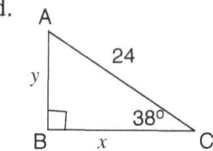

e.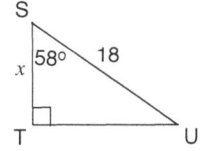

f.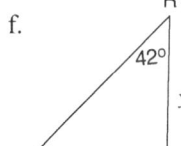

22. For Exercises **a-f**, find the value of each variable.
≪See Example 8-18≫

a.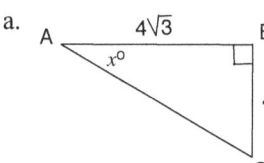
b.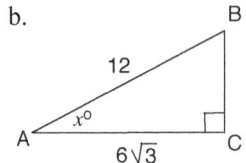
c.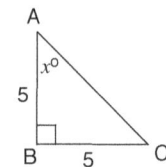

d.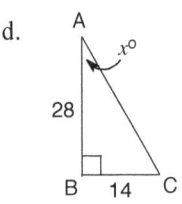
e.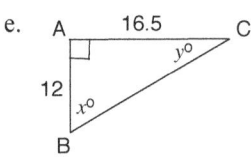
f.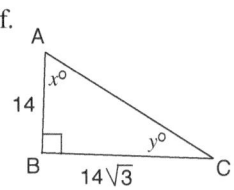

23. For Exercises **a-f**, find the trigonometric ratios.
≪See Examples 8-14 and 8-15≫

a. sin A
   sin C

b. cos A
   cos B

c. tan B
   tan C

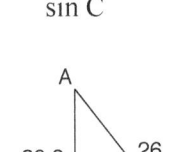
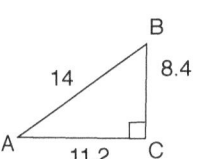
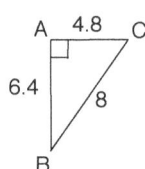

d. sin A
   sin C

e. cos A
   cos B

f. tan A
   tan C

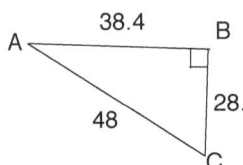
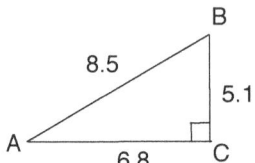
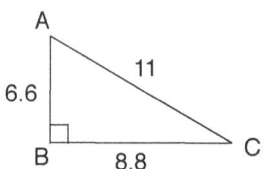

24. Find the sine, cosine, and tangent of angle A and B.
《See Examples 8-14 and 8-15》

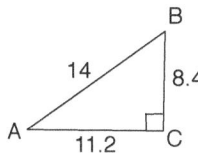

a. sin A   b. sin B   c. cos A

d. cos B   e. tan A   f. tan B

25. Find the sine, cosine, and tangent of ∠A and ∠B.
《See Examples 8-14 and 8-15》

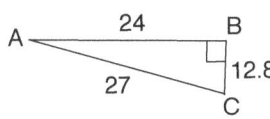

a. sin A   b. sin C   c. cos A

d. cos C   e. tan A   f. tan C

26. Find $m\angle A$ and $m\angle C$ and round to the nearest degree if necessary.
《See Examples 8-17 and 8-18》

a. 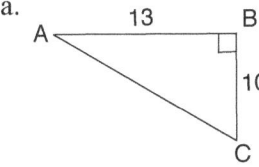   b. 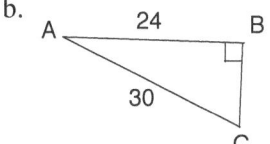   c.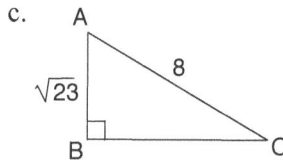

27. For Exercises **a-f**, find the values of each variable.
《See Examples 8-15 to 8-18》

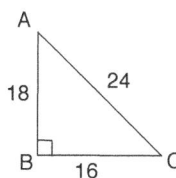

a. sin A   b. sin C   c. cos A

d. cos C   e. tan A   f. tan C

28. Use a calculator to find each angle measure to the nearest tenth of a degree.
≪See Example 8-16≫

   a. $\sin^{-1}(0.33)$ b. $\sin^{-1}\left(\frac{3}{5}\right)$ c. $\sin^{-1}(1.33)$

   d. $\cos^{-1}(0.85)$ e. $\tan^{-1}\left(\frac{17}{8}\right)$ f. $\tan^{-1}\left(\frac{8}{7}\right)$

   g. $\sin^{-1}(0.62)$ h. $\sin^{-1}(0.56)$ i. $\cos^{-1}\left(\frac{9}{5}\right)$

29. For Exercises **a-f**, find the measures of ∠A and ∠B. Round decimals to the nearest tenth.
≪See Examples 8-17 and 8-18≫

a. 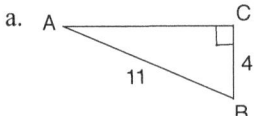   b. 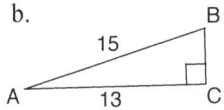   c.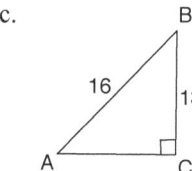

d. 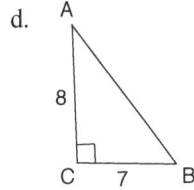   e. 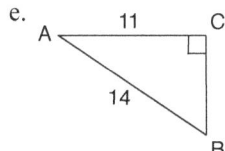   f.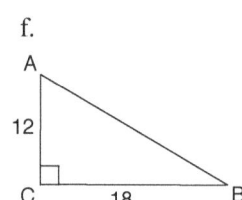

30. For Exercises **a-f**, find the unknown values. Round decimals to the nearest hundredth.
≪See Examples 8-17 and 8-18≫

a. Find $m\angle A$ and $m\angle B$.

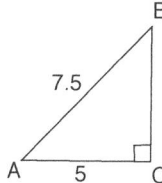

b. Find $m\angle T$ and $m\angle U$.

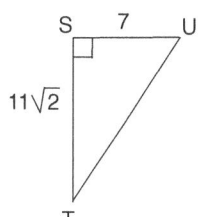

c. Find the lengths of AB and BC.

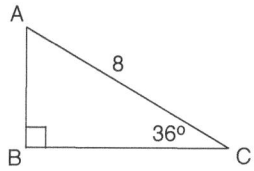

d. Find the lengths of OQ and QR.

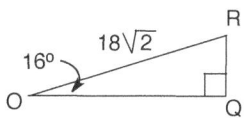

e. Find the values of x and y.

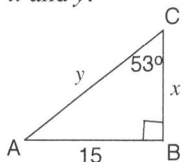

f. Find the length of DE.

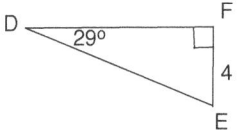

31. For Exercises **a-d**, find the unknown measures. Round decimals to the nearest tenth.
≪See Example 8-18≫

a.

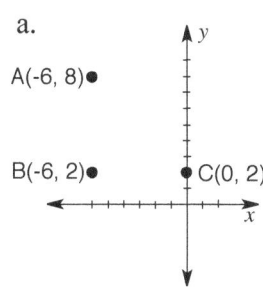

b.

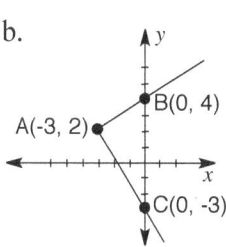

c.

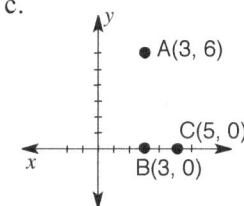

d.

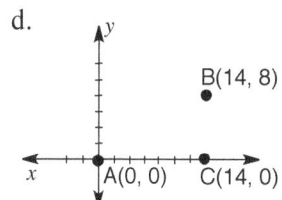

32. In Exercises **a-f**, use the diagram to find the unknown values. Round decimals to the nearest tenth.
≪See Examples 8-17 and 8-18≫

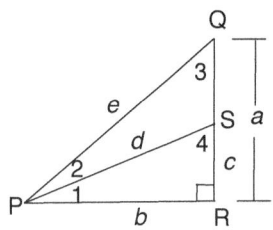

a. If $b = 17$ cm and $c = 8$ cm, find d, $m\angle 1$ and $m\angle 4$.

b. If $c = 10$ cm and $m\angle 1 = 32°$, find the lengths of b and d.

c. If $a = 25$ cm and $b = 21$ cm, find the measures of e and $m\angle 3$.

d. If $e = 32$ cm and $b = 20$ cm, find sine Q, cosine Q, and tangent Q.

e. If $b = 11$ cm and $m\angle 4 = 58°$, find the lengths of d and c.

f. If $b = 13$ cm and $c = 5$ cm, find d, sine 1, cosine 1, and tangent 1.

33. In Exercises **a-d**, use the diagram to find the unknown measure. Round lengths to the nearest tenth.
≪See Examples 8-17 and 8-18≫

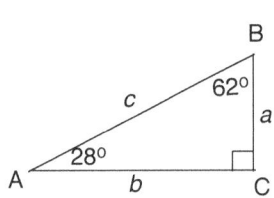

a. If $a = 6$ ft, find the length of b.

b. If $b = 15$ in., find the length of c.

c. If $c = 29$ cm, find the lengths of a and b.

d. If $b = 14$ cm, find the lengths of a and c.

34. Describe each vector as an ordered pair.
≪See Example 8-19≫

a.
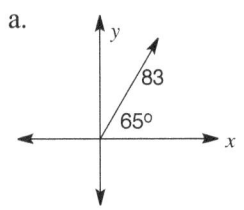

b.
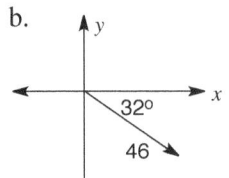

c.
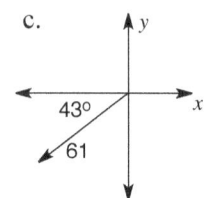

35. Use compass directions to find the distance and the direction of each vector.
≪See Example 8-19≫

a. 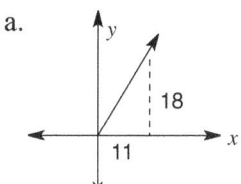   b. 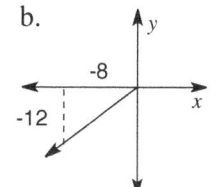   c.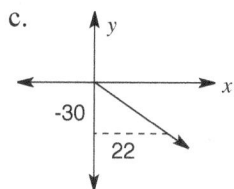

36. Find the magnitude and direction of each vector.
≪See Example 8-20≫

a. $\overrightarrow{AB}$: A(3, 4), B(5, 3)   b. $\overrightarrow{PQ}$: P(−4, 4), Q(−1, 5)

c. $\overrightarrow{ST}$: S(2, −5), T(−1, −4)   d. $\overrightarrow{AB}$: A(−3, 2), B(−3, −5)

e. $\overrightarrow{PQ}$: P(−2, −2), Q(−3, 3)   f. $\overrightarrow{ST}$: S(−4, 2), T(5, 5)

37. For Exercises **a-f**, write the sum of the two vectors as an ordered pair and draw the resulting vector.
≪See Examples 8-19 and 8-20≫

a. 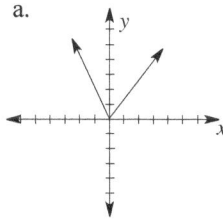   b. 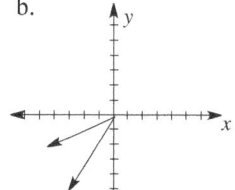   c.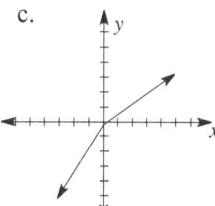

d. 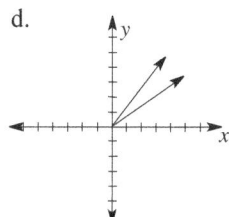   e. 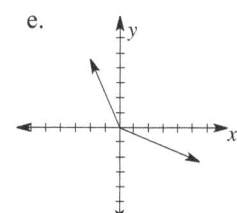   f.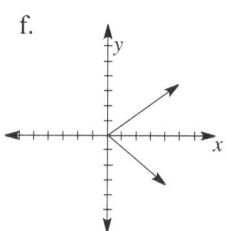

ANSWERS

(1) a. 6, b. $13\frac{1}{3}$, c. $16\frac{2}{3}$, d. $10\frac{2}{3}$

(2) a. $19\frac{3}{5}$, b. 46.3, c. $\sqrt{30}$, d. $3\sqrt{3}$, e. $10\frac{2}{3}$, f. 9.8, g. 54, h. $5\frac{4}{9}$, i. $13\frac{4}{9}$

(3) a. 5, b. 6, c. $3\sqrt{6}$, d. 12.75, e. $6\sqrt{7}$, f. 12.5

(4) a. 9, b. $9\frac{3}{13}$, c. $16\frac{9}{22}$, d. 41.1, e. $y = 11$, f. $2\sqrt{42}$, g. $x = 9, y = 16, z = 20$, h. 7.875, i. $x = 14.4, y = 25.6, z = 24, h = 19.2$

(5) a. 9, b. 18, c. $3\sqrt{10}$

(6) a. $\sqrt{34}$, b. $\sqrt{106}$, c. $4\sqrt{5}$, d. 5, e. $\sqrt{74}$, f. $6\sqrt{5}$, g. 15, h. 2.5, i. $2\sqrt{5}$

(7) a. 48 units^2, b. 13.5 units^2, c. 13.6 units^2

(8) a. $y = x = 8\sqrt{2}$ m, b. 64 m^2, c. $32\sqrt{2}$ m

(9) a. 7.2, b. 51.3 m^2, c. 43 m

(10) a. $2\sqrt{91}$, b. 20, c. obtuse, d. acute, e. obtuse

(11) c = 12 units, A = 72 units^2,

(12) a. right triangle, b. right triangle, c. right triangle, d. obtuse, e. right triangle, f. obtuse, g. acute, h. right triangle, i. obtuse

(13) a. acute, b. acute, c. right triangle, d. acute, e. acute, f. obtuse, g. acute, h. right triangle,

(14) a. $7\sqrt{2}$, b. 10, c. $13\sqrt{2}$, d. 18, e. $x = 3\sqrt{2}, y = 6$, f. $x = y = \frac{5\sqrt{10}}{2}$,

(15) c = 9.9 cm, A = 49 cm^2

(16) a. 4, b. 12, c. $4\sqrt{6}$, d. $2\sqrt{6}$, e. $x = 12, y = 12\sqrt{3}$, f. $y = 7\sqrt{5}, x = 7\sqrt{15}$

(17) a. b = 9, d = $9\sqrt{3}$, b. a = 14, c = $7\sqrt{2}$, c. e = $6\sqrt{2}$, d = $6\sqrt{6}$, d. a = 16, b = 8, e. a = 12, c = $6\sqrt{2}$, f. d = $7\sqrt{3}$, e = 7,

(18) a. $x = \frac{9\sqrt{6}}{2}$, b. $x = y = 6\sqrt{3}$, c. $y = 9, x = 9\sqrt{3}$, d. 30, e. $x = 7\sqrt{6}, y = 21\sqrt{2}$, f. $11\sqrt{6}$

(19) a. 60°, b. 60°, c. 60°, d. 30°, e. 30°, f. 30°, g. 45°, h. 45°, i. 45°, j. 45°, k. 45°, l. 45°

(20) a. 32.8 cm, b. 36 cm, c. $(8\sqrt{3})/3$, d. 24

(21) a. 6.76, b. $y = 5.9, x = 9.5$, c. $x = 4.7, y = 7.6$, d. $x = 22, y = 14.8$, e. 9.5, f. 11.1

(22) a. 30°, b. 30°, c. 45°, d. 26.6°, e. $y = 36°, x = 54°$, f. $y = 30°, x = 60°$

(23) a. $\sin A = \frac{15.6}{26}, \sin C = \frac{20.8}{26}$, b. $\cos A = \frac{11.2}{14}$, $\cos B = \frac{8.4}{14}$, c. $\tan B = \frac{4.8}{6.4}, \tan C = \frac{6.4}{4.8}$, d. $\sin A = \frac{28}{48}, \sin C = \frac{38.4}{48}$, e. $\cos A = \frac{6.8}{8.5}, \cos B = \frac{5.1}{8.5}$, f. $\tan A = \frac{8.8}{6.6}, \tan C = \frac{6.6}{8.8}$

(24) a. 8.4/14, 36.9°, b. 11.2/14, 53.1°, c. 11.2/14, 36.9°, d. 8.4/14, 53.1°, e. 8.4/11.2, 36.9°, f. 11.2/8.4, 53.1°

(25) a. 0.31, 18°, b. 0.95, 72°, c. 0.95, 18°, d. 0.31, 72°, e. 0.325, 18°, f. 3.08, 72°

(26) a. $m\angle A = 38°, m\angle C = 52°$, b. $m\angle A = 37°, m\angle C = 53°$, c. $m\angle A = 53°, m\angle C = 37°$

(27) a. 0.666, 41.8°, b. 0.75, 48.6°, c. 0.75, 41.4°, d. 0.666, 48.2°, e. 0.888, 41.6°, f. 1.125, 48.4°

(28) a. 19.3°, b. 36.9°, c. 53.1°, d. 31.8°, e. 65°, f. 48.8°, g. 38.5°, h. 55.9°, i. 60.9°

(29) a. $m\angle A = 21.3°, m\angle B = 68.7°$, b. $m\angle A = 29.9°, m\angle B = 60.1°$, c. $m\angle A = 54.3°, m\angle B = 63.3°$, d. $m\angle A = 41.2°, m\angle B = 47.8°$, e. $m\angle A = 38.2°, m\angle B = 51.8°$, f. $m\angle A = 56.3°, m\angle B = 33.7°$

(30) a. $m\angle A = 48.2°, m\angle B = 41.8°$, b. $m\angle T = 30°, m\angle U = 60°$, c. AB = 4.7, BC = 6.5, d. OQ = 24.5, QR = 7, e. $x = 11.3, y = 18.8$, f. DE = 8.3

(31) a. $m\angle A = 45°, m\angle C = 45°$, b. $m\angle B = 59°, m\angle C = 31°$, c. $m\angle A = 18.4°, m\angle B = 71.6°$, d. $m\angle A = 29.7°, m\angle B = 60.3°$

(32) a. d = 18.8 units, $m\angle 1 = 25.2°, m\angle 4 = 64.8°$, b. b = 16 cm, d = 18.9 cm, c. e = 32.6, $m\angle 3 = 40°$, d. sin Q = 20/32, cos Q = 25/32, tan Q = 20/25, e. d = 13 cm, c = 6.9 cm, f. d = 14 cm, sin 1 = 5/14, cos 1 = 13/14, tan 1 = 5/13,

(33) a. b = 11.3 ft, b. c = 17 in., c. a = 13.6 cm, b = 25.6 cm, d. a = 7.4 cm, c = 15.9 cm

(34) a. <51.5, 83>, b. <44.1, 46>, c. <90, 61>

(35) a. 58.6°, b. 265.9°, c. 306.3°

(36)
a. 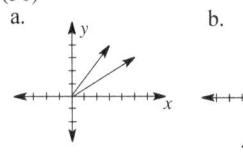 b.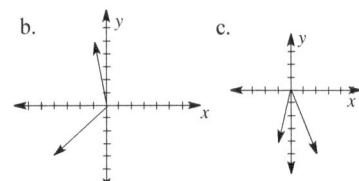
c.
d. 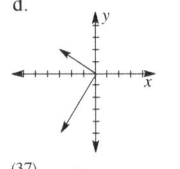 e. 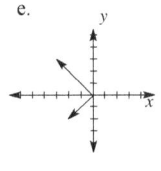 f.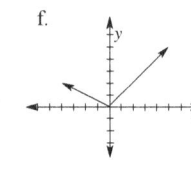

(37)
a. 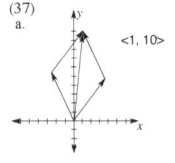 <1, 10> b. 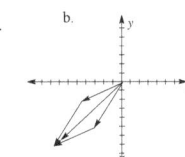 c. 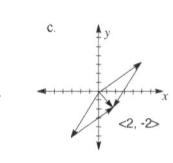 <2, -2>
d. 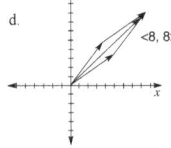 <8, 8> e. <5.5, 2.5> f. 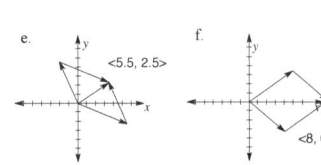 <8, 0>

SELF-TEST

1. Which of the following statements defines the similar right triangles as shown below?

   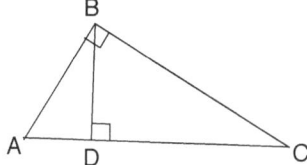

   (a) △ABD ~ △ABC
   (b) △CDB ~ △ABC
   (c) △ABD ~ △CDB
   (d) All of the above.

2. Given the diagram from Question **1**, which of the following statements is true?

   (a) BD is the geometric mean of AD and CD.
   (b) AB is the geometric mean of AC and AD.
   (c) BC is the geometric mean of AC and CD.
   (d) All of the above.

3. Given the diagram from Question **1**, which of the following statement is not true?

   (a) $\dfrac{CD}{BD} = \dfrac{BD}{AC}$
   (b) $\dfrac{AB}{AC} = \dfrac{AC}{BD}$
   (c) $\dfrac{AC}{BC} = \dfrac{BC}{CD}$
   (d) $\dfrac{AC}{AB} = \dfrac{AB}{AD}$

4. Given the diagram from Question **1**, △ABD and △CBD are similar right triangles that are separated by the altitude BD. If the length of the altitude BD is 15 ft and the length of AD is 8 ft, find the length of DC. Round to the nearest tenth.

   (a) 14.0 ft
   (b) 28.1 ft
   (c) 17.0 ft
   (d) 15.0 ft

5. △ABD and △CBD are similar right triangles that are separated by the altitude BD. If the length of BC is 18 and the length of DC is 14, find the length of AC. Round to the nearest tenth.

   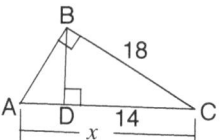

   (a) 19.6
   (b) 11.3
   (c) 9.1
   (d) 23.1

6. △ACD and △BCD are similar right triangles that are separated by the altitude CD. The length of AD is three times longer than BD. If the length of BD is 6, find the length of DC. Round to the nearest tenth.

   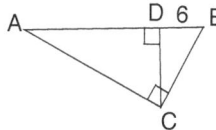

   (a) 10.4
   (b) 18.0
   (c) 9.0
   (d) 6.7

7. In the diagram, △ABD and △CBD are similar right triangles that are separated by the altitude BD. If the length of AB is 10.5 units and the length of AC is 13.5 units, find the length of AD. Round to the nearest tenth.

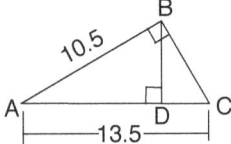

(a) 8.5
(b) 8.2
(c) 7.9
(d) 7.6

8. In the diagram, △ABD and △CBD are similar right triangles that are separated by the altitude BD. Find the length of the altitude BD. Round to the nearest tenth.

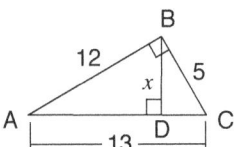

(a) 4.5
(b) 4.6
(c) 4.7
(d) 4.8

9. △ABC and △CBD are similar right triangles that are separated by the altitude BD. Find the value of x. Round your answer to the nearest tenth.

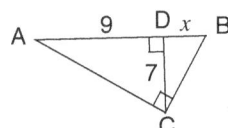

(a) 5.0
(b) 5.2
(c) 5.4
(d) 5.6

10. Which of the following are the values of x and y? Round your answer to the nearest tenth.

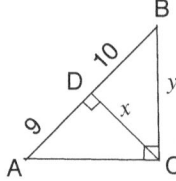

(a) $x = 9.5$ and $y = 12.8$
(b) $x = 9.5$ and $y = 13.8$
(c) $x = 8.5$ and $y = 12.8$
(d) $x = 8.5$ and $y = 13.8$

11. Which of the following is the geometric mean of 6 and 18?

(a) $\frac{1}{2}(6^2 \times 18^2)$
(b) $\sqrt{(6)(18)}$
(c) $\frac{1}{2}\sqrt{(6)(18)}$
(d) $\frac{1}{2}(6 \times 18)$

12. Which of the following is the geometric mean of 3 and 8? Round to the nearest tenth.

(a) 5.5
(b) 11.0
(c) 8.5
(d) 4.9

13. Given that 5 is a geometric mean of $\sqrt{35}$, what is the other geometric mean?

(a) 7
(b) 8
(c) 9
(d) 10

14. Which of the following statement defines the Pythagorean Theorem?

(a) $a^2 = c^2 + b^2$, where c is the length of the hypotenuse and a and b are the lengths of the legs.
(b) $c^2 = a^2 - b^2$, where c is the length of the hypotenuse and a and b are the lengths of the legs.
(c) $c^2 = a^2 + b^2$, where c is the length of the hypotenuse and a and b are the lengths of the legs.
(d) $c^2 = (a + b)(a + b)$, where c is the length of the hypotenuse and a and b are the lengths of the legs.

15. Find the measure of ∠A. Round to the nearest whole number.

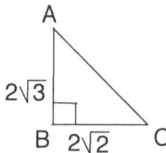

(a) 39.2°
(b) 50.8°
(c) 54.7°
(d) 35.3°

16. Which equation is true for the right triangle below?

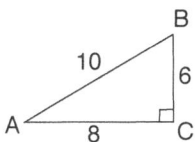

(a) $8^2 = 10^2 - 6^2$
(b) $10^2 = \sqrt{(8)(6)}$
(c) $10^2 = 8^2 + 6^2$
(d) both (a) and (c)

17. Use the Pythagorean Theorem to find the length of the shorter leg of a given triangle if the length of the longer leg is 25 ft and the length of the hypotenuse is 34 ft. Round your answer to the nearest tenth.

(a) 9.0
(b) 18.0
(c) 23.0
(d) 29.2

18. Which of the following is the third length for a given right triangle that has two side lengths of 12 and 20?

(a) 15.5
(b) 16
(c) 23.3
(d) 17

19. Find the measure of ∠A. Round to the nearest whole number.

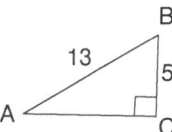

(a) 23°
(b) 24°
(c) 35°
(d) 69°

20. Which of the following statement defines the Pythagorean Theorem?

(a) If $c^2 = a^2 + b^2$, then it is a right triangle, where c is the length of the hypotenuse and a and b are the length of the legs.
(b) If $c^2 > a^2 + b^2$, then it is an obtuse triangle, where c is the length of the hypotenuse and a and b are the length of the legs.
(c) $c^2 < a^2 + b^2$, then it is an acute triangle, where c is the length of the hypotenuse and a and b are the length of the legs.
(d) $c^2 < a^2 + b^2$, then it is an obtuse triangle, where c is the length of the hypotenuse and a and b are the length of the legs.

21. Find the measure of ∠A. Round to the nearest whole number.

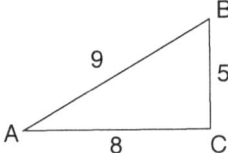

(a) 58°
(b) 22°
(c) 48°
(d) 32°

22. The side length of a given square is $10\sqrt{2}$ units. Find the length of diagonal a.

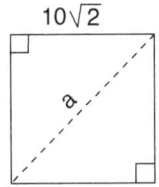

(a) 20
(b) 21
(c) 12
(d) 15

23. Classify the triangle given the lengths of its sides are 10, $4\sqrt{2}$, and $8\sqrt{2}$.

(a) acute
(b) right
(c) obtuse
(d) none of the above

24. Classify the triangle given the lengths of its sides are 15, 7, and 17.

(a) acute
(b) right
(c) obtuse
(d) none of the above

25. Which of the following statements does not define special right triangles?

(a) In a 45°-45°-90° triangle, the hypotenuse is $\sqrt{2}$ times as long as each leg.
(b) In a 45°-45°-90° triangle, if the length of the hypotenuse is 3 cm, the length of each leg is $3\sqrt{2}$ cm.
(c) In a 30°-60°-90° triangle, the hypotenuse is twice as long as the shorter leg, and the longer leg is $\sqrt{3}$ times as long as the shorter leg.
(d) In a 30°-60°-90° triangle, if the length of the hypotenuse is 3 cm, the lengths of each leg are $\frac{3\sqrt{3}}{2}$ and $\frac{3}{2}$ cm.

26. In the diagram below, ΔABC is a 45°-45°-90° triangle. Find the value of x.

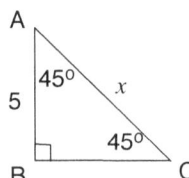

(a) 8
(b) $5\sqrt{3}$
(c) 10
(d) $5\sqrt{2}$

27. In the diagram below, ΔABC is a 30°-60°-90° triangle. Find the value of x.

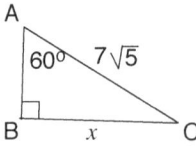

(a) 7
(b) $\frac{14}{2}\sqrt{15}$
(c) $\frac{7}{2}\sqrt{5}$
(d) $\frac{7}{2}\sqrt{15}$

28. In the diagram below, ΔABC is an isosceles right triangle. Which of the following is the length of BC?

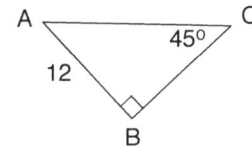

(a) $12\sqrt{3}$
(b) $12\sqrt{2}$
(c) 12
(d) 6

29. In a 45°-45°-90° triangle, ΔABC is an isosceles right triangle. Which of the following is the length of the hypotenuse if the length of a leg is $7\sqrt{2}$?

(a) 14
(b) $7\sqrt{2}$
(c) 7
(d) $14\sqrt{2}$

30. In a 30°-60°-90° triangle, which of the following are the lengths of the longer leg (LL) and the hypotenuse (H) if the length of a shorter leg is 6?

(a) LL = $6\sqrt{2}$ and H = 12
(b) LL = $6\sqrt{3}$ and H = 12
(c) LL = 12 and H = 12
(d) LL = $6\sqrt{3}$ and H = $6\sqrt{3}$

31. ΔABC is an isosceles triangle. Which of the following is the length of BC?

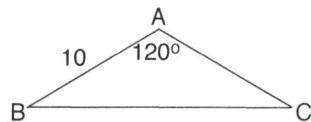

(a) $10\sqrt{3}$
(b) $10\sqrt{2}$
(c) 20
(d) $20\sqrt{3}$

32. In the diagram below, ΔABC is an isosceles right triangle. Which of the following is the length of BC?

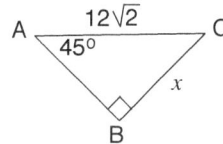

(a) 12
(b) 10
(c) 11
(d) $12\sqrt{6}$

33. In a 30°-60°-90° triangle, which of the following is the length of the hypotenuse?

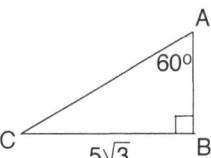

(a) 5
(b) $5\sqrt{6}$
(c) $\frac{5}{2}\sqrt{6}$
(d) $5\sqrt{6}$

34. In a 30°-60°-90° triangle, which of the following are the lengths of the hypotenuse and BC?

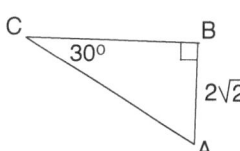

(a) $4\sqrt{6}$
(b) $2\sqrt{6}$
(c) 2
(d) $3\sqrt{2}$

35. △ABC is a 30°-60°-90° triangle. Find the value of x.

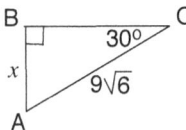

(a) $\frac{9}{2}\sqrt{2}$
(b) 9
(c) $\frac{9}{2}\sqrt{6}$
(d) $9\sqrt{2}$

36. Which of the following statement is not a definition of the trigonometric ratios in the diagram?

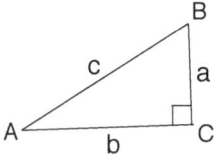

(a) $\cos A = \frac{b}{c}$
(b) $\cos B = \frac{a}{c}$
(c) $\sin A = \frac{a}{c}$
(d) $\tan A = \frac{b}{a}$

37. Given the diagram from Question 36, which of the following statement is not a definition of the trigonometric ratios given the diagram?

(a) $\tan B = \frac{b}{a}$
(b) $\sin B = \frac{a}{c}$
(c) $\tan A = \frac{a}{b}$
(d) $\sin A = \frac{a}{c}$

38. Find the value of x. Round your answer to the nearest tenth.

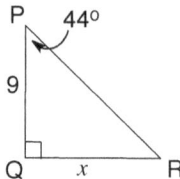

(a) 13.0
(b) 12.5
(c) 9.3
(d) 8.7

39. Find the value of x. Round your answer to the nearest tenth.

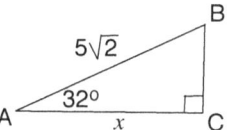

(a) 6.0
(b) 3.7
(c) 13.3
(d) 8.3

40. Find the value of y. Round your answer to the nearest tenth.

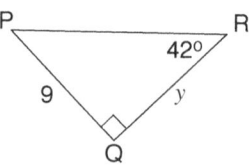

(a) 10
(b) 8.1
(c) 13.5
(d) 6.0

41. Which of the following is the ratio for tan 30°?

(a) $\frac{7\sqrt{3}}{8}$
(b) $\frac{4}{7\sqrt{3}}$
(c) $\frac{8}{4}$
(d) $\frac{4}{8}$

42. Which of the following is the ratio for tan 40°?

(a) $\frac{5}{6}$
(b) $\frac{6}{5}$
(c) $\frac{7.8}{5}$
(d) $\frac{6}{7.8}$

43. Which of the following is the angle measure if the tangent value is 0.96? Round to the nearest tenth.

(a) 71.8°
(b) 46.5°
(c) 18.2°
(d) 43.5°

44. Which of the following is the angle measure if the value of cosine is $\frac{\sqrt{3}}{6}$? Round to the nearest tenth.

(a) 16.1°
(b) 16.8°
(c) 73.2°
(d) 73.9°

45. Use the diagram to find the measure of ∠A. Round to the nearest whole number.

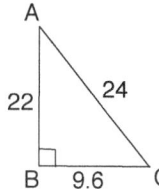

(a) 16.1°
(b) 16.8°
(c) 23.6°
(d) 66.4°

46. Find the measure of ∠A. Round to the nearest whole number.

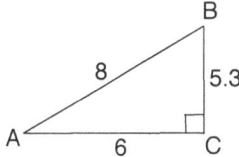

(a) 41.4°
(b) 48.6°
(c) 33.5°
(d) 56.5°

47. Which of the following is the angle measure if sin⁻¹ (0.36)? Round to the nearest tenth.

(a) 21.1°
(b) 19.8°
(c) 68.9°
(d) 70.2°

48. Which of the following is the angle measure if tan⁻¹ ($\frac{15}{12}$)? Round to the nearest tenth.

(a) 16.1°
(b) 38.7°
(c) 51.3°
(d) 36.9°

49. In the diagram below, find the measure of ∠C. Round to the nearest tenth.

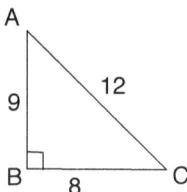

(a) 48.4°
(b) 41.6°
(c) 48.1°
(d) 41.9°

50. James is standing 32 ft away from a tree. He measures a 48° angle from his standing point to the top of the tree. How tall is the tree? Round your answer to the nearest tenth.

(a) 23.7 ft
(b) 28.8 ft
(c) 35.5 ft
(d) 43.1 ft

51. Olivia is flying a kite. Her kite is 42 ft up in the air with 70 ft of the string unraveled and she is standing 56 ft away from where the kite is. What is the angle of the kite to her standing point of the ground?

(a) 16.1°
(b) 38.7°
(c) 51.3°
(d) 36.9°

ANSWERS

(1) d	(2) d	(3) b	(4) b	(5) d	(6) a
(7) b	(8) c	(9) c	(10) b	(11) b	(12) d
(13) a	(14) c	(15) a	(16) d	(17) c	(18) b
(19) a	(20) d	(21) d	(22) a	(23) c	(24) a
(25) b	(26) d	(27) d	(28) c	(29) a	(30) b
(31) a	(32) a	(33) a	(34) b	(35) c	(36) d
(37) b	(38) d	(39) a	(40) a	(41) a	(42) a
(43) d	(44) c	(45) c	(46) a	(47) a	(48) c
(49) a	(50) c	(51) d			

CHAPTER 9
Circles

In this chapter, you will identify segments and lines related to circles, use properties of tangents, arcs, and chords of a circle as well as the properties of inscribed angles and inscribed polygons of a circle, and finding and graphing the equation of a circle. You will also be able to find the length of tangents, chords, and other lines that intersect a circle.

[CONCEPTS]　[EXAMPLES]　[FORMULAS]　[VOCABULARY]

9-1. List of terms for the lines and points of a circle.

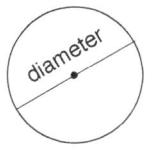 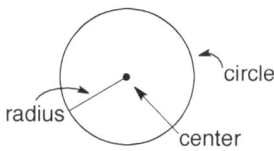 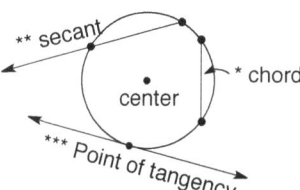

　　* chord: a segment whose endpoints are on a circle.
　　** secant: a line that intersects a circle in two points.
　　*** tangent: a line that intersects a circle in exactly one point.

9-2. In the diagram below, BP = 17 cm and the radius of the circle is 8 cm. AB is tangent to the circle at P. What is the length of AB?

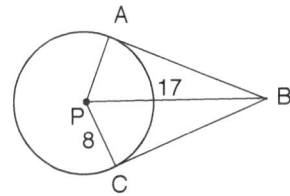

SOLUTION

In the given information, AB is tangent to the circle at P, which makes it perpendicular to AP. Since AP and CP are both radii to circle at P, AP = CP. Thus, the segments form a right triangle and therefore, you can use the Pythagorean Theorem to find AB.

$BP^2 = AP^2 + AB^2$　　　　Use the Pythagorean Theorem
$AB^2 = BP^2 - AP^2$　　　　Subtract AP^2 from both sides.
$AB^2 = 17^2 - 8^2$　　　　　Substitute.
$(AB)^2 = 225$　　　　　　　Simplify.
$AB = 15$　　　　　　　　　Square root

9-3. In the diagram, AB is tangent to the circle P. Find $m\angle APB$.

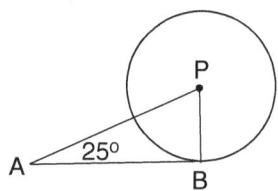

SOLUTION

AB is tangent to the circle at P in the given information. So AB is perpendicular to PB. Therefore, $m\angle ABP$ is 90°.

$180° = 90° + 25° + m\angle APB$ Triangle Sum Theorem
$m\angle APB = 65°$ Simplify.

9-4. Identify each line segment in the diagram below.

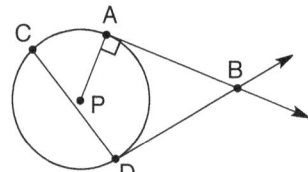

a. AP

b. AB

c. CD

SOLUTION

a. AP is a radius on circle P.
b. AB is the line of tangent on circle P.
c. CD is a chord on circle P.

9-5. In the diagram, △APC is an isosceles triangle. AC = 16 cm and the radius of the circle PB is 6 cm. AC is tangent to the circle at P. What is the length of AD?

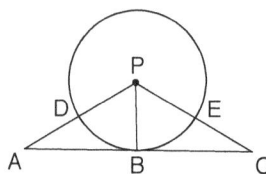

SOLUTION

The length of BP is 6 cm because it is the radius of the circle.
Let the length of AP represent x.

$x^2 = AB^2 + BP^2$	Pythagorean Theorem
$x^2 = 8^2 + 6^2$	Substitute.
$x^2 = 64 + 36$	Simplify.
$x^2 = 100$	Simplify.
$x = AP = 10$	Square root.
$AD = AP - PD$	$x = AP$ and $PD =$ radius
$AD = 10 - 6$	Substitute.
$AD = 4$	Add 6 to both sides.

9-6. Find the measure of each arc in circle P and the name of each arc.

a. $\widehat{AC}$

b. $\widehat{AD}$

c. $\widehat{ADB}$

d. $\widehat{ABC}$

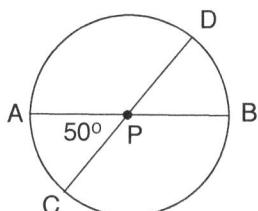

SOLUTION

If the measure of a central angle is less than 180°, it is called a <u>minor arc</u>. If the measure of a central angle is between 180° and 360°, it is called a <u>major arc</u>. If the measure of a central angle is exactly 180°, it is called a <u>semicircle</u>.

a. $\widehat{AC}$ and $\angle APC$ are equal. So $m\widehat{AC} = 50°$. $\widehat{AC}$ is a minor arc.
b. The measure of $\widehat{AD}$ with $\widehat{AC}$ is a semicircle. $m\widehat{AD} + m\widehat{AC} = 180°$. $m\widehat{AD} = 180° - m\widehat{AC}$. $m\widehat{AD} = 180° - 50°$. So $m\widehat{AD} = 130°$. $\widehat{AD}$ is a minor arc.
c. $m\widehat{ADB} = 180°$. $\widehat{ADB}$ is a semicircle.
d. $m\widehat{ABC} = 360° - 50°$. So $m\widehat{ABC} = 310°$. $\widehat{ABC}$ is a major arc.

9-7. Find the value of x and y.

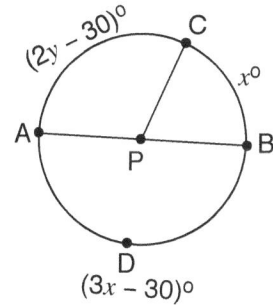

SOLUTION

The measure of a semicircle is 180°. So because $m\widehat{ADB} = 3x - 30$, by substitution, $180 = 3x - 30$. Find the value of x. As $\widehat{AC}$ and $\widehat{CB}$ form a semicircle, the sum of their measures is 180°. Therefore, $2y - 30 + x = 180$. Insert the value in the equation so that there is only one variable to solve for.

$\widehat{ADB}$ is a semicircle in circle P.	$\widehat{AC}$ and $\widehat{BC}$ forms another semicircle in circle
So $m\widehat{ADB} = 180°$.	$m\widehat{AC} + m\widehat{BC} = 180°$
$(3x - 30)° = 180°$	$(2y - 30)° + x° = 180°$
$3x = 210$	$2y - 30 + 70 = 180$
$x = 70$	$2y + 40 = 180$
	$2y = 140$
	$y = 70$

9-8. In the diagram, if BD = 18 and EF = 12, find the length of the radius of circle P and round your answer to the nearest hundredth.

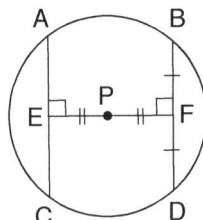

SOLUTION

Connect C to P to form line CP. CP is a radius of circle P as it extends from the center to the circumference of the circle. △CEP is a right triangle because FE is the perpendicular bisector of AC so ∠CEP is a right angle. Use the Pythagorean Theorem to find the length of CP.

△CEP is a right triangle. BD = AC = 18 and EC = 9, EF = 12 and EP = 6.
$CP^2 = EC^2 + EP^2$ Pythagorean Theorem
$CP^2 = 9^2 + 6^2$ Substitute.
$CP^2 = 117$ Simplify.
$CP = 10.82$ Square root
So the radius in the circle is about 10.82.

9-9. In the diagram, find the length of BC if the radius is 7 cm in the circle P and $m\angle BPC$ is 120°.

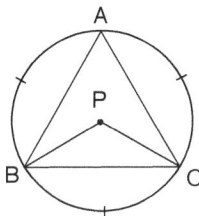

SOLUTION

ΔBPC is an isosceles triangle and the center P lies on the perpendicular line on BC. So $m\angle BCP$ is 30° as well $m\angle CBP$. The triangle (ΔCDP) below is a 30°-60°-90° right triangle.

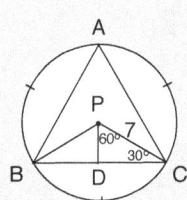

hypotenuse = (2)(shorter leg)
PC = 2(PD)
7 = (2)(PD)
3.5 = PD

longer leg = $\sqrt{3}$ • shorter leg
CD = ($\sqrt{3}$)(PD)
CD = $\sqrt{3}$(3.5)
Now you can find the length of BC.
BC = 2(CD)
BC = 2(3.5$\sqrt{3}$)
BC = 7$\sqrt{3}$

9-10. Find the values of x and y.

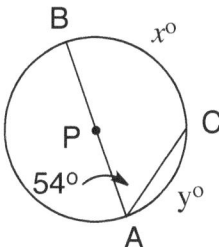

SOLUTION

By the definition of the measure of an inscribed angle, if an angle is inscribed in a circle, then its measure is half the measure of its intercepted arc. Given that $m\widehat{BC} = x$ and $m\widehat{AC} = y$, as $\angle BAC$ intercepts $\widehat{BC}$, multiply the measure of $\angle BAC$ by 2 to find the measure of the intercepted arc.
Because $\widehat{BC}$ and $\widehat{AC}$ form a semicircle, the sum of their measures is 180°. So solve for y.

$\boxed{m\angle BAC = \tfrac{1}{2}m\widehat{BC}}$

2($m\angle BAC$) = $m\widehat{BC}$ Multiply 2 on both sides.
    2(54°) = $m\widehat{BC}$ Substitute.
    108° = $m\widehat{BC}$ Simplify.
So, $x = m\widehat{BC} = 108°$

$\overline{AB}$ is a diameter in a circle P and has
$m\widehat{BC} + m\widehat{AC} = 180°$
$m\widehat{AC} = 180° - m\widehat{BC}$
$m\widehat{AC} = 180° - 108°$
$m\widehat{AC} = 72°$
So, $y = m\widehat{AC} = 72°$.

9-11. Find the values of each variable.

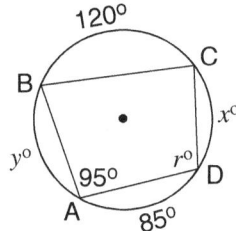

SOLUTION

By the definition of the measure of an inscribed angle, if an angle is inscribed in a circle, then its measure is half the measure of the intercepted arc.

$\boxed{m\angle BAD = \tfrac{1}{2}m\widehat{BCD}}$

2($m\angle BAD$) = $m\widehat{BCD}$
    2(95°) = $m\widehat{BCD}$
    190° = $m\widehat{BCD}$

To find the value of x.
$m\widehat{BCD} = m\widehat{BC} + m\widehat{DC}$
$m\widehat{BCD} - m\widehat{BC} = m\widehat{DC}$
190° − 120° = $m\widehat{DC}$
70° = $m\widehat{DC}$
So, $x = m\widehat{DC} = 70°$.

To find the value of y.
360° − ($m\widehat{BC} + m\widehat{DC} + m\widehat{AD}$) = $m\widehat{AB}$
360° − (120° + 70° + 85°) = $m\widehat{AB}$
360° − 275° = $m\widehat{AB}$
85° = $m\widehat{AB}$
So, $y = m\widehat{AB} = 85°$

To find the value of r.
$r = \tfrac{1}{2}(m\widehat{AB} + m\widehat{BC})$
$r = \tfrac{1}{2}(85° + 120°)$
$r = \tfrac{1}{2}(205°)$
$r = 102.5°$

So the values of x, y, and r are 70°, 85°, and 102.5° respectively.

9-12. Find $m\angle C$ and $m\angle D$ if $m\widehat{AB}$ is 120°.

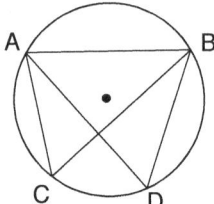

Chapter 9   Circles   75

SOLUTION

By the definition of an inscribed angle, if two inscribed angles of a circle intercept the same arc, then the angles are congruent. Both ∠C and ∠D are the intercepted angles of $\widehat{AB}$. So to find $m\angle C$ and $m\angle D$, divide $\widehat{AB}$ by two.

$m\angle C = \frac{1}{2}m\widehat{AB}$ $m\angle D = \frac{1}{2}m\widehat{AB}$

$m\angle C = \frac{1}{2}(120°)$ $m\angle D = \frac{1}{2}(120°)$

$m\angle C = 60°$ $m\angle D = 60°$

So the measures of ∠C and ∠D are both 60°.

9-13. Find the value of x.

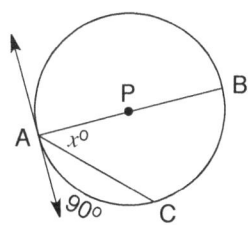

SOLUTION

$\widehat{ACB}$ is a semicircle and $\widehat{AC}$ is a part of $\widehat{ACB}$. To find $\widehat{BC}$, subtract the measure of $\widehat{AC}$ from 180°. Then divide the measure of $\widehat{BC}$ by two to find x.

$\widehat{BC} + \widehat{AC} = 180°$ $m\angle BAC = \frac{1}{2}m\widehat{BC}$

$\widehat{BC} = 180° - \widehat{AC}$ $m\angle BAC = \frac{1}{2}(90°)$

$\widehat{BC} = 180° - 90°$ $m\angle BAC = 45°$

$\widehat{BC} = 90°$

So the value of $x°$ is 45°.

9-14. Find the value of x.

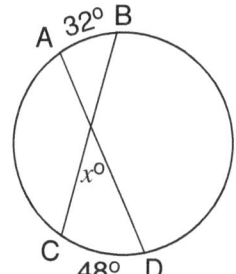

SOLUTION

To find the measure of an angle whose vertex is in the interior of the circle, note that the measure of x is equal to half of the sum of the measures of $\widehat{AB}$ and $\widehat{CD}$.

$$m\angle x = \tfrac{1}{2}(m\widehat{AB} + m\widehat{CD})$$

$x = \tfrac{1}{2}(m\widehat{AB} + m\widehat{CD})$

$x = \tfrac{1}{2}(32° + 48°)$

$x = \tfrac{1}{2}(80°)$

$x = 40°$

So the value of x is 40°.

9-15. Find the value of x.

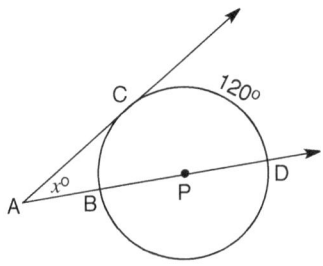

SOLUTION

AC is tangent to the circle P and BD is a diameter of the circle. You can find the measure and intercepted arcs. If a <u>secant</u> intersects in the exterior of a circle, then the measure of the angle formed is one half times the difference of the measure of the intercepted arcs. $m\angle A = \tfrac{1}{2}(m\widehat{CD} - m\widehat{BC})$

First, find $\widehat{BC}$

$m\widehat{BC} = 180° - 120°$

$m\widehat{BC} = 60°$

Now you can find the value of x.

$$m\angle x = \tfrac{1}{2}(m\widehat{CD} - m\widehat{BC})$$

$x = \tfrac{1}{2}(m\widehat{CD} - m\widehat{BC})$

$x = \tfrac{1}{2}(120° - 60°)$

$x = \tfrac{1}{2}(60°)$

$x = 30°$

So the value of $x°$ is 30°.

9-16. In the diagram, the measures of $\widehat{AB}$, $\widehat{AC}$, and $\widehat{CD}$ are congruent. What is the measure of $\angle V$ if the measure of $\widehat{ACD}$ is 210°?

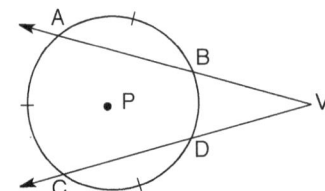

> **SOLUTION**
>
> Remember that $m\widehat{AC}$ is equal to $m\widehat{CD}$ in the given information. So $m\widehat{AC} = 105°$. AC is tangent to the circle P and BD is a diameter. If two secants intersect in the exterior of the circle, then the measure of the angle formed is one half the product of the measure of the intercepted arcs. So you can now find the measure of the intercepted arcs.
>
> $m\angle V = \frac{1}{2}(m\widehat{AC} - m\widehat{BD})$
>
> $m\widehat{ACD} = 210°$
>
> $m\widehat{AC} = m\widehat{CD} = m\widehat{AB} = 105°$ or $m\widehat{AC} + m\widehat{CD} + m\widehat{AB} = 315°$
>
> Now find the measure of $\widehat{BD}$.
>
> $m\widehat{BD} = 360° - 315° = 45°$
>
> $\boxed{m\angle V = \frac{1}{2}(m\widehat{AC} - m\widehat{BD})}$
>
> $m\angle V = \frac{1}{2}(105° - 45°)$
>
> $m\angle V = \frac{1}{2}(60°)$
>
> $m\angle V = 30°$

9-17. Find the values of the variable.

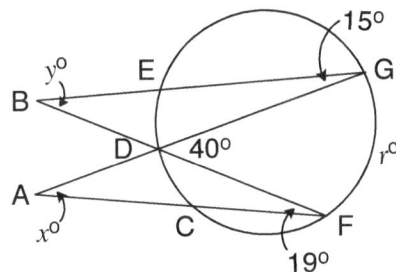

> **SOLUTION**
>
> There are three different angles ($\angle FDG$, $\angle GDF$, and $\angle CFD$) that are inscribed in a circle. To find the measure of their intercepted arcs, use the known inscribed angles such as $m\angle DGE = (1/2)m\widehat{DE}$, $m\angle FDG = (1/2) m\widehat{GF}$, and $m\angle CFD = (1/2)m\widehat{CD}$. After that, by using two different angles of each two secants intersect in the exterior of a circle, find the measures of the angle formed by one half the difference of the measures of the intercepted arcs.
>
> $r = \widehat{FG} = 2(m\angle FDG) = 2(40°)$
>
> $r = 80°$
>
> $m\widehat{CD} = 2(m\angle CFD) = 2(19°)$ $m\widehat{DE} = 2(m\angle DGE) = 2(15°)$
>
> $m\widehat{CD} = 38°$ $m\widehat{DE} = 30°$
>
> To find the value of x.   To find the value of y.

$$\boxed{m\angle x = \tfrac{1}{2}(m\widehat{FG} - m\widehat{CD})}$$
$x = \tfrac{1}{2}(m\widehat{FG} - m\widehat{CD})$
$x = \tfrac{1}{2}(80° - 38°)$
$x = \tfrac{1}{2}(42°)$
$x = 21°$

$$\boxed{m\angle y = \tfrac{1}{2}(m\widehat{FG} - m\widehat{DE})}$$
$y = \tfrac{1}{2}(m\widehat{FG} - m\widehat{DE})$
$y = \tfrac{1}{2}(80° - 30°)$
$y = \tfrac{1}{2}(50°)$
$y = 25°$

So the values of x and y are 21° and 25° respectively.

9-18. Find the value of x.

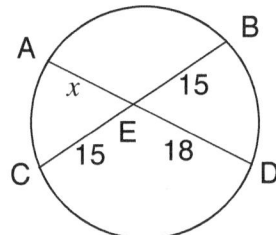

SOLUTION

You can find two chords that intersect inside the circle. Chords AD and BC intersect in the interior of circle E. So the product of AE and DE is equal to the product of CE and BE.

$$\boxed{(AE)(DE) = (CE)(BE)}$$

$x(18) = (15)(15)$
$18x = 225$
$x \approx 12.5$

So the value of x is about 12.5.

9-19. Find the value of x.

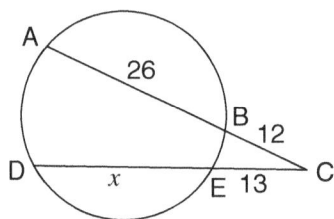

SOLUTION

Two secant segments AB and DE share the same endpoint outside a circle. So the product of the lengths of BC and AC equals the product of the lengths of CE and DC.

$$\boxed{(BC)(AC) = (CE)(DC)}$$

$(12)(26 + 12) = (13)(x + 13)$
$(12)(38) = 13x + 169$
$456 = 13x + 169$
$287 = 13x$
$22.1 \approx x$

So the value of x is about 22.1.

9-20. Find the value of x.

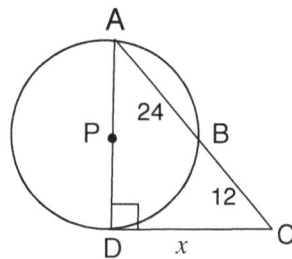

SOLUTION

Tangent segment CD and secant segment AC share an endpoint outside of the circle at C. So the product of the lengths of BC and AC equals the length of DC^2.

$$(BC)(AC) = DC^2$$

$(12)(12 + 24) = (x)^2$
$432 = x^2$
$20.8 \approx x$

9-21. Write the standard equation of each circle that the given information.
(notations: center = c, radius = r, diameter = d)
$c = (-2, -5)$ and $r = 3$

SOLUTION

The standard equation of a circle can be used as the same as the Distance Formula.

$$r^2 = (x_1 - x_0)^2 + (y_1 - y_0)^2 \quad \begin{array}{l}\text{center} = (x_0, y_0)\\ \text{any point on the circle} = (x_1, y_1)\end{array}$$

$3^2 = (x_1 - (-2))^2 + (y_1 - (-5))^2$
$9 = (x_1 + 2)^2 + (y_1 + 5)^2$

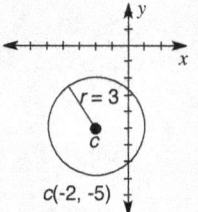

9-22. What are the center and the radius of a circle given that its equation is $x^2 + y^2 - 10x + 24 = 0$?

SOLUTION

$$r^2 = (x_1 - x_0)^2 + (y_1 - y_0)^2 \quad \begin{array}{l}\text{center} = (x_0, y_0)\\ \text{any point on the circle} = (x_1, y_1)\end{array}$$

$x^2 - 10x + y^2 + 24 = 0$	Given
$x^2 - 10x + y^2 = -24$	Rearrange the equation.
$x^2 - 10x + 25 + y^2 = -24 + 25$	Add 25 to both sides.
$(x - 5)(x - 5) + y^2 = 1$	Factor the polynomial.
$(x - 5)^2 + y^2 = 1$	Write the equation of the circle.

The center in the circle is $(5, 0)$.
The radius in the circle is 1.

PRACTICES

1. In the exercises below, identify whether the line or segment is a chord, a secant, a tangent, a radius, or a diameter of circle P.
   ≪See Example 9-1≫

   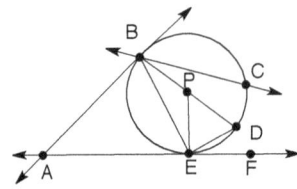

   a. $\overline{AB}$ b. $\overline{BC}$ c. $\overline{EP}$

   d. $\overline{BD}$ e. $\overline{DE}$ f. $\overline{AE}$

   g. $\overline{BE}$ h. $\overline{EF}$

2. For Exercises **a-h**, identify the arcs of circle P.
   ≪See Example 9-6≫

   I. minor arc   II. major arc   III. Semicircle

   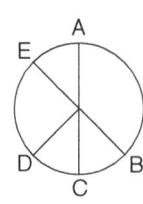

   a. $\widehat{AE}$ b. $\widehat{ABC}$ c. $\widehat{AB}$

   d. $\widehat{DE}$ e. $\widehat{BAE}$ f. $\widehat{ADB}$

   g. $\widehat{CD}$ h. $\widehat{ABD}$

3. For Exercises **a-f**, determine the values of the unknown variables of circle P. If necessary, round your answer to the nearest tenth.
   ≪See Examples 9-2 to 9-5≫

Chapter 9   Circles   81

a. $m\angle APC$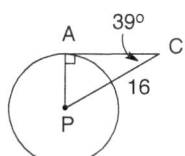

b. $\overline{AP}$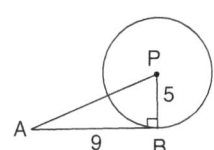

c. Find r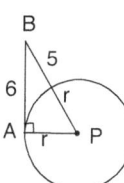

d. Find x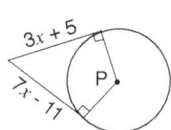

e. Find r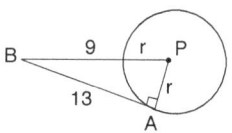

f. Find BP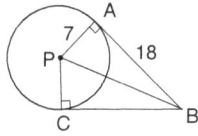

4. For Exercises **a-c**, determine the values of the unknown variables of circle P. If necessary, round your answer to the nearest tenth.
《See Examples 9-2 to 9-5》

a. Find x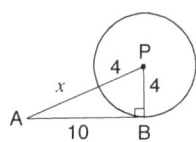

b. Find $\overline{AC}$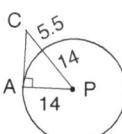

c. Find $\overline{AB}$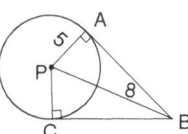

5. For Exercises **a-e**, determine the values of the unknown variables of circle P. If necessary, round your answer to the nearest tenth.
《See Examples 9-2 to 9-5》

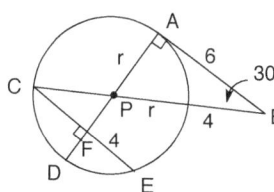

a. $\overline{AP}$

b. $m\angle APB$

c. $m\angle CPF$

d. $m\angle FCP$

e. $\overline{FP}$

6. For Exercises **a-f**, determine the values of the unknown variables of circle P. If necessary, round your answer to the nearest tenth.
《See Examples 9-2 to 9-5》

a.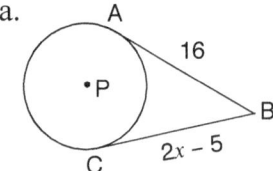

b.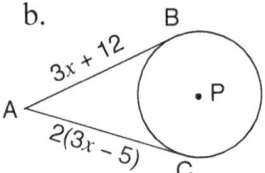

c.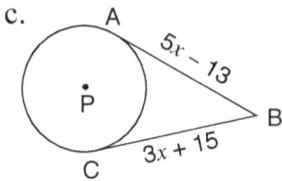

d.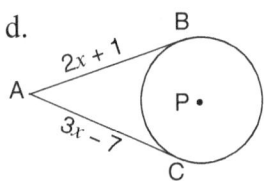

e.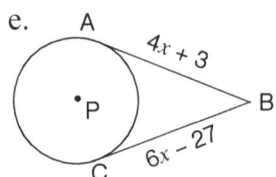

f.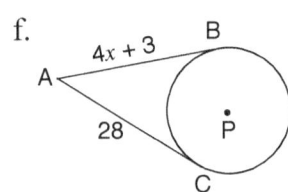

7. Find each arc measure in circle P. AE is perpendicular to CG.
《See Examples 9-6 and 9-7》

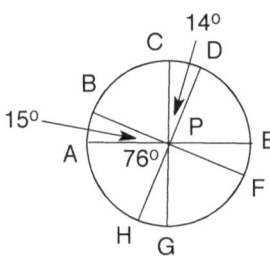

a. $\overset{\frown}{BC}$ b. $\overset{\frown}{CDE}$ c. $\overset{\frown}{DE}$

d. $\overset{\frown}{EF}$ e. $\overset{\frown}{BCE}$ f. $\overset{\frown}{AHG}$

g. $\overset{\frown}{FG}$ h. $\overset{\frown}{EFG}$ i. $\overset{\frown}{GH}$

j. $\overset{\frown}{FGH}$ k. $\overset{\frown}{EFH}$ l. $\overset{\frown}{BD}$

8. Find the value of x.
≪See Examples 9-6 and 9-7≫

a.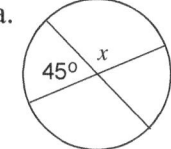

b.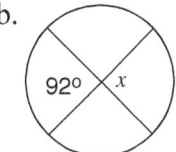

c.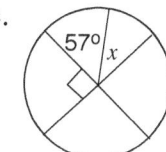

d.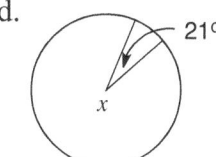

e.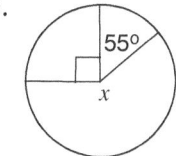

f.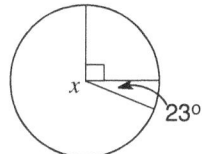

9. For Exercises **a-f**, find the value of x.
≪See Examples 9-6 and 9-7≫

a.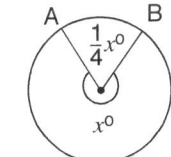

b.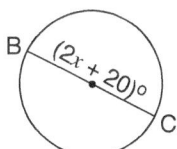

c.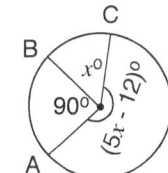

d.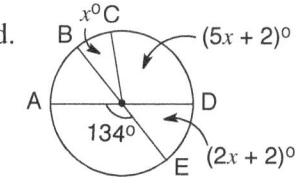

e.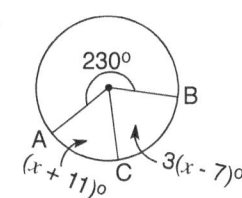

f.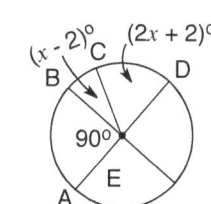

10. For Exercises **a-f**, find the value of x. Round to the nearest tenth if necessary.
≪See Example 9-8≫

a.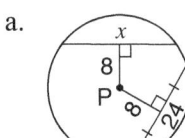

b.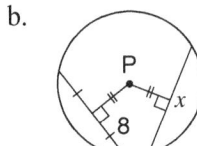

c.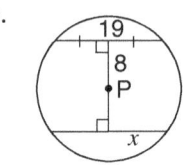

d.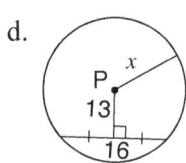

e.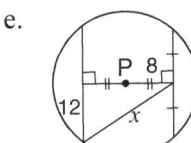

f.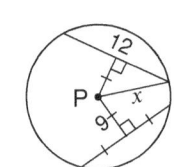

11. For Exercises **a-c** below, determine the measure of $\widehat{QR}$ using trigonometry.
≪See Example 9-8≫

a.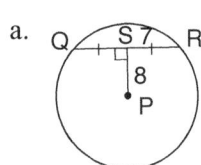

b.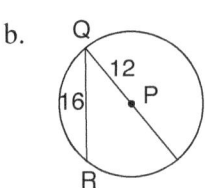

c.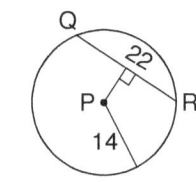

12. For Exercises **a-f**, find the value of each variable.
≪See Examples 9-6 to 9-8≫

a.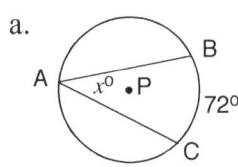

b.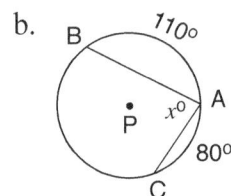

c.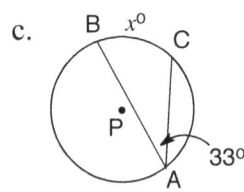

d.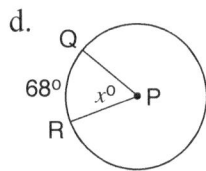

e.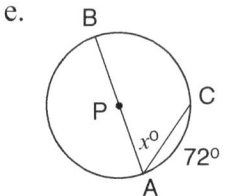

f.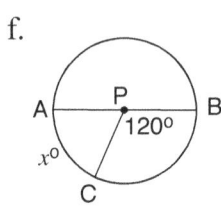

13. Find each value.
«See Example 9-8»

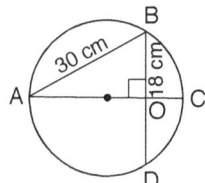

a. $\overline{AO}$
b. $m\angle BAO$
c. $m\widehat{AB}$
d. $m\widehat{BC}$
e. $m\widehat{AD}$
f. $m\widehat{DC}$

14. For Exercises **a-f**, find the value of each variable.
«See Examples 9-6 and 9-12»

a. 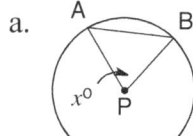
b. 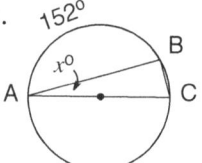
c.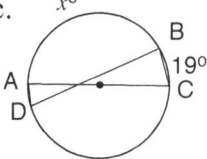

d. 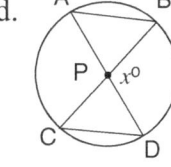
e. 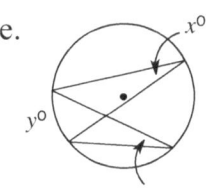
f.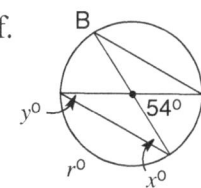

15. For Exercises **a-i**, find the value of each variable.
«See Examples 9-11 and 9-12»

a. 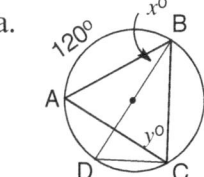
b. 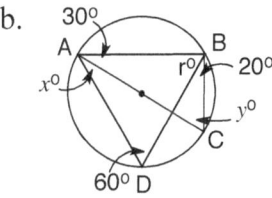
c.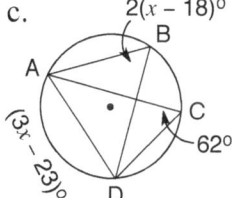

d.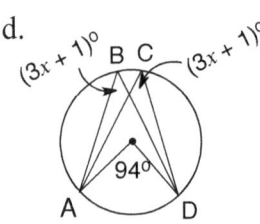

e.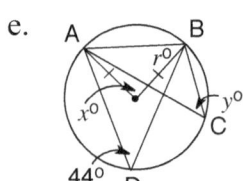

f.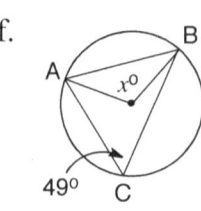

g.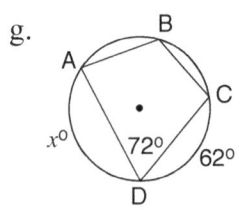

h.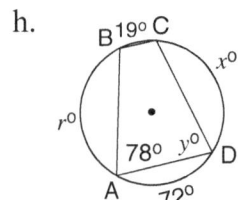

i.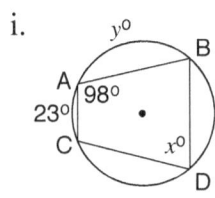

16. Find the value of x.
≪See Example 9-12≫

a.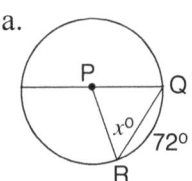

b.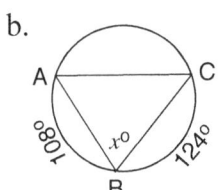

c.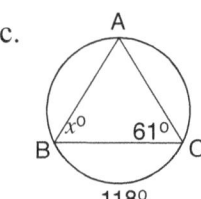

17. For Exercises **a-f**, find the value of each variable.
≪See Example 9-13≫

a.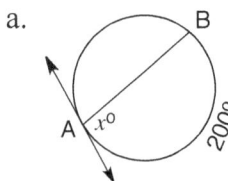

b.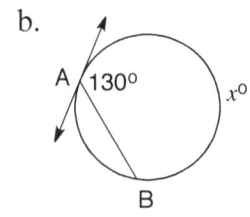

c.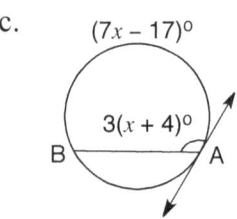

d.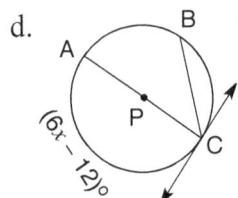

e.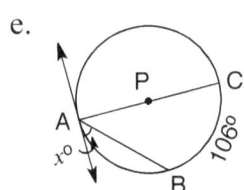

f.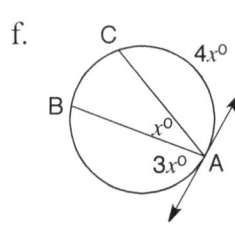

18. For Exercises **a-f**, find the value of each variable.
《See Example 9-14》

a.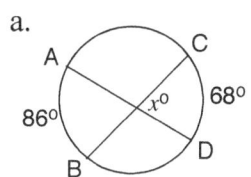
b.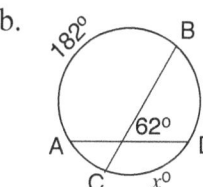
c.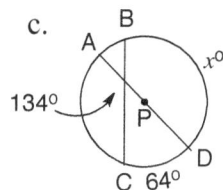

d.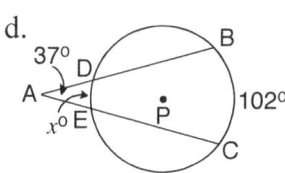
e.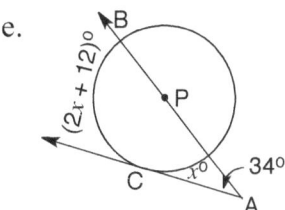
f.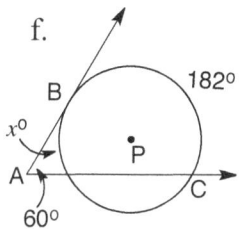

19. Find each angle or arc measure.
《See Example 9-15》

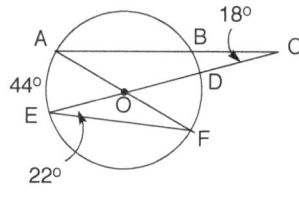

a. $m\widehat{BD}$ b. $m\widehat{DF}$

c. $m\widehat{AB}$ d. $m\angle AOD$

e. $m\angle AOE$ f. $m\angle DOF$

g. $m\widehat{EF}$ h. $m\angle EOF$

i. $m\angle EFO$ j. $m\angle BAF$

20. For Exercises **a-f**, find the values of the variables.
《See Examples 9-13 to 9-16》

a.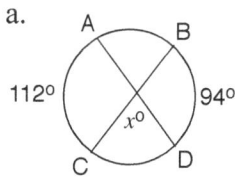
b.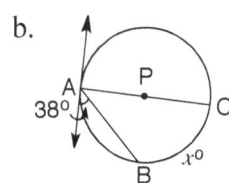
c.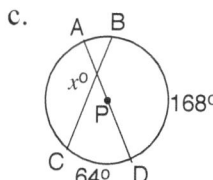

d.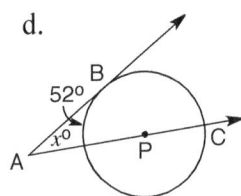
e.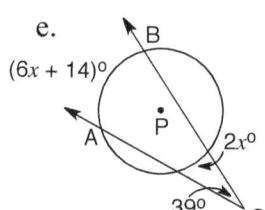
f.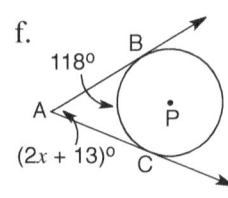

21. Find each angle or arc measure.
≪See Examples 9-16 and 9-17≫

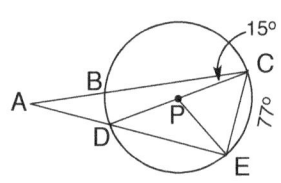

a. $m\widehat{DE}$
b. $m\widehat{DB}$
c. $m\widehat{BC}$
d. $m\angle DPE$
e. $m\angle BAD$
f. $m\angle EDP$
g. $m\angle DEP$
h. $m\angle CPE$
i. $m\angle ECP$
j. $m\angle CEP$

22. For Exercises **a-f**, find the values of the variables.
≪See Examples 9-16 and 9-17≫

a.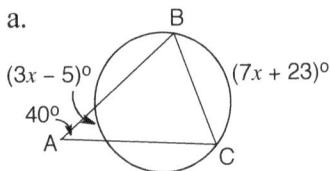
b.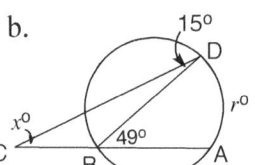
c.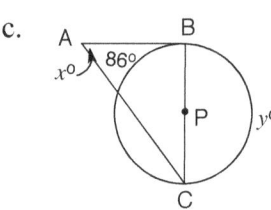

d.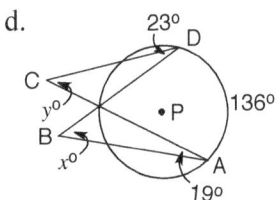
e.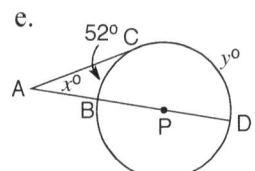
f.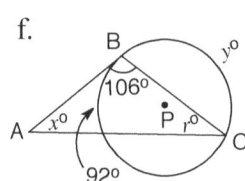

23. For Exercises **a-f**, find the values of the variables.
≪See Example 9-18≫

a.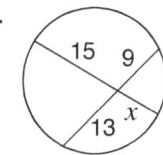

b.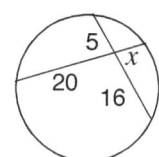

c.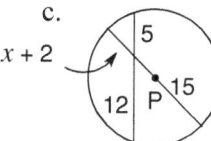

d.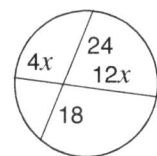

e.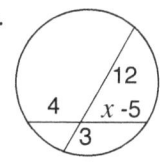

f.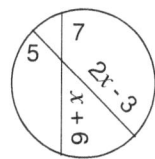

24. Find each arc or angle measure.
≪See Example 9-19≫

a.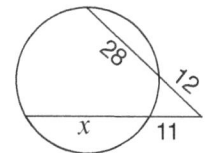

b.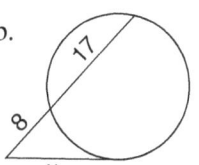

c.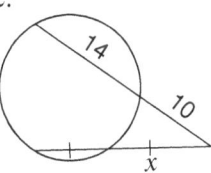

d.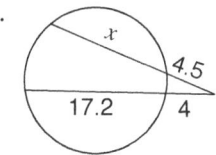

e.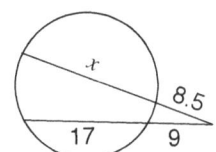

f.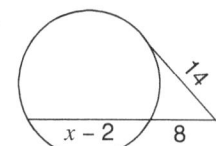

25. For Exercises **a-n**, the radius of circle P is 10 units and BC bisects AD. Find the value of each variable.
≪See Example 9-18≫

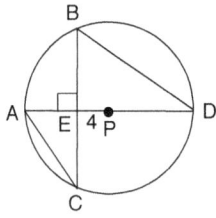

a. $\overline{AE}$	b. $\overline{CE}$	c. $\overline{AC}$
d. $\overline{BE}$	e. $\overline{DE}$	f. $\overline{BD}$
h. $m\angle EAC$	i. $m\angle ACE$	j. $m\widehat{AC}$
k. $m\widehat{AB}$	l. $m\widehat{CD}$	n. $m\widehat{BD}$

26. For Exercises **a-c**, find the value of x.
《See Examples 9-18 and 9-19》

a. 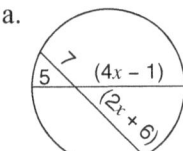 b. 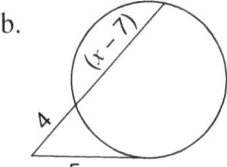 c.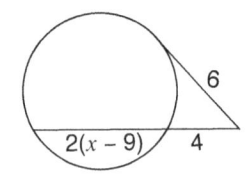

27. Find each arc or angle measure.
《See Example 9-20》

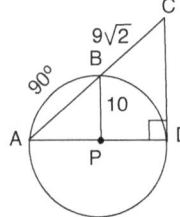

a. $\overline{AP}$	b. $\overline{AB}$
c. $\overline{CD}$	d. $m\widehat{AD}$
e. $m\widehat{BD}$	f. $m\angle ACD$
g. $m\angle CAD$	i. $\overline{DP}$

28. Find each angle or arc measure.
≪See Example 9-20≫

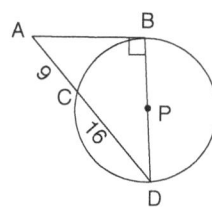

a. $\overline{AB}$ b. $\overline{BD}$ c. $m\angle BAD$

d. $m\widehat{BC}$ e. $m\angle BDC$ f. $m\widehat{CD}$

29. Write the standard equation of each circle using the given information.
(notations: center = c, radius = r, diameter = d)
≪See Example 9-21≫

a. $c = (3, 3)$ and $r = 2$ b. $c = (-4, 2)$ and $r = 5$

c. $c = (-2, -3)$ and $r = 3$ d. $c = (5, -2)$ and $r = 1$

e. $c = (-3, 0)$ and $d = 4$ f. $c = (-3, -5)$ and $r = 7$

g. $c = (5, 4)$ and $d = 2$ h. $c = (2, 0)$ and $r = 4$

i. $c = (4, -1)$ and $d = 6$ j. $c = (-5, -5)$ and $r = 2$

k. $c = (0, -3)$ and $r = 7$ l. $c = (0, 0)$ and $r = 2$

30. Find the coordinates for the center and radius of the circle by using the standard equation.
≪See Example 9-22≫

a. $(x + 4)^2 + (y + 3)^2 = 1$ b. $(x - 3)^2 + (y + 3)^2 = 2$

c. $(x - 3)^2 + (y + 5)^2 = 49$ d. $(x + 3)^2 + (y)^2 = 9$

e. $(x + A)^2 + (y - B)^2 = C$ f. $(x - 7)^2 + (y - 6)^2 = 10$

g. $(x + 3)^2 + (y + 2)^2 = 25$ h. $(x - 4)^2 + (y - 3)^2 = 100$

i. $(x - 5)^2 + (y)^2 = 1$ j. $(x)^2 + (y)^2 = 49$

31. Find the quadrant of the coordinates of the center of a circle using the given equation.
《See Examples 9-21 and 9-22》

a. $(x + 5)^2 + (y - 7)^2 = 2$

b. $(x + 3)^2 + (y + 5)^2 = 4$

c. $(x - 4)^2 + (y - 6)^2 = 9$

d. $(x - A)^2 + (y + B)^2 = C^2$

32. Find the best answer to each given problem.
《See Examples 9-21 and 9-22》

a. Which of the following is the equation of a circle with a radius of 4 and its center at P(−3, 2)?

A. $(x + 3)^2 + (y - 2)^2 = 4$
B. $(x + 3)^2 + (y - 2)^2 = 16$
C. $(x - 3)^2 + (y + 2)^2 - 16 = 0$
D. $(x - 3)^2 + (y - 2)^2 - 4 = 0$

b. Which of the following is the equation of a circle with a radius of 10 and its center at P(0, 2)?

A. $(x)^2 + (y - 2)^2 = 10$
B. $(x)^2 + (y + 2)^2 = 10$
C. $(x)^2 + (y - 2)^2 = 100$
D. $(x)^2 + (y + 2)^2 = 100$

c. Which of the following is the equation of a circle with a diameter of 10 and its center at P(4, −2)?

A. $(x - 4)^2 + (y + 2)^2 = 10$
B. $(x - 4)^2 + (y + 2)^2 = 100$
C. $(x - 4)^2 + (y + 2)^2 = 5$
D. $(x - 4)^2 + (y + 2)^2 = 25$

d. Which of the following is the equation of a circle with a radius of 1 and its center at (6, 0)?

A. $(x + 6)^2 + (y)^2 - 1 = 0$
B. $(x + 6)^2 + (y)^2 + 1 = 0$
C. $(x - 6)^2 + (y)^2 - 1 = 0$
D. $(x - 6)^2 + (y)^2 + 1 = 0$

e. What is the radius of the circle with an equation of $(x - 4)^2 + (y)^2 - 25 = 0$?

A. 25
B. 4
C. 5
D. 2

f. What are the coordinates of the center of the circle indicated by the equation $(x + 4)^2 + (y - 9)^2 = 36$?

A. (4, 9)
B. (−4, −9)
C. (−4, 9)
D. (4, −9)

g. Which equation of the circle is located in the second quadrant?

A. $(x + 3)^2 + (y - 2)^2 - 1 = 0$
B. $(x + 6)^2 + (y + 2)^2 + 1 = 0$
C. $(x - 3)^2 + (y + 2)^2 - 1 = 0$
D. $(x - 6)^2 + (y - 2)^2 + 1 = 0$

h. The equation of a circle is $(x - 1)^2 + (y + 1)^2 = 1$. What quadrant is it located?

A. I
B. II
C. III
D. IV

i. What are the center and radius of the circle indicated by the equation $(x)^2 + (y - 1)^2 = 4$?

A. center (0, −1) and radius of 4
B. center (0, 1) and radius of 4
C. center (0, −1) and radius of 2
D. center (0, 1) and radius of 2

ANSWERS

(1) a. tangent, b. secant, c. radius, d. diameter, e. chord, f. tangent, g. chord, h. tangent

(2) a. minor, b. semicircle, c. minor, d. minor, e. semicircle, f. major, g. minor, h. major

(3) a. 51°, b. 10.3, c. 1.1, d. 4, e. 4.9, f. 19.3

(4) a. 6.8, b. 13.6, c. 12

(5) a. 2.5, b. 60°, c. 60°, d. 30°, e. 2.5

(6) a. 10.5, b. 7.3, c. 14, d. 8, e. 15, f. 6.25

(7) a. 75°, b. 90°, c. 76°, d. 15°, e. 175°, f. 90°, g. 75°, h. 90°, i. 14°, j. 90°, k. 76°, l. 90°

(8) a. 135°, b. 92°, c. 33°, d. 339°, e. 215°, f. 247°

(9) a. 288°, b. 80°, c. $x = 47.2$, d. 22°, e. 35°, f. 60°

(10) a. 24, b. 16, c. 9.5, d. 15.3, e. 20, f. 10.8

(11) a. 82.4°, b. 83.6°, c. 79°

(12) a. 36°, b. 85°, c. 66°, d. 68°, e. 54°, f. 60°

(13) a. 24 cm, b. 36.9°, c. 106.2°, d. 73.8°, e. 143°, f. 106.2°, g. 73.8°

(14) a. 70°, b. 28°, c. 161°, d. 125°, e. $x = 27$, $y = 54°$, f. $x = 27°$, $y = 27°$, $r = 126°$

(15) a. $x = 30°$, $y = 60°$, b. $x = 20°$, $y = 60°$, $r = 70°$, c. $x = 49$, d. $x = 31$, e. $x = 88°$, $y = 44°$, $r = 46$, f. $x = 98°$, g. $x = 154°$, h. $x = 137°$, $y = 75.5$, $r = 132$, i. $x = 82°$, $y = 141$

(16) a. 54°, b. 64°, c. 60°

(17) a. 100°, b. 260°, c. $x = 41$, d. $x = 32$, e. 37°, f. 30°

(18) a. 77°, b. 54°, c. 152°, d. 28°, e. 56°, f. 62°

(19) a. 8°, b. 44°, c. 128°, d. 136°, e. 44°, f. 44°, g. 136°, h. 136°, i. 22°, j. 26°

(20) a. 77°, b. 104°, c. 115°, d. 38°, e. 16°, f. 24.5

(21) a. 103°, b. 30°, c. 150°, d. 103°, e. 23.5°, f. 38.5°, g. 38.5°, h. 77°, i. 51.5°, j. 51.5°

(22) a. 15.5, b. $x = 34$, $r = 98$, c. $x = 47$, $y = 180$, d. $x = 49$, $y = 45$, e. $x = 38$, $y = 128$, f. $x = 28$, $y = 148$, $r = 46$

(23) a. 7.8, b. 4, c. 2, d. 3, e. 14, f. 19

(24) a. 32.6, b. 14.1, c. 8.4, d. 14.6, e. 10, f. 18.5

(25) a. 6, b. $\sqrt{84}$, c. $\sqrt{120}$, d. 9.165, e. 14, f. 16.7, g. 56.8°, h. 33.2°, i. 66.4°, j. 66.4°, k. 113.5°, l. 113.5°

(26) a. 7.8, b. 9.25, c. 11.5

(27) a. 10, b. 14.14, c. 18, d. 180°, e. 90°, f. 45°, g. 45°, i. 10

(28) a. 15, b. 20, c. 53.13°, d. 73.74°, e. 36.87°, f. 106.26°

(29) a. $(x - 3)^2 + (y - 3)^2 = 4$, b. $(x + 4)^2 + (y - 2)^2 = 25$, c. $(x + 2)^2 + (y + 3)^2 = 9$, d. $(x - 5)^2 + (y + 2)^2 = 1$, e. $(x + 3)^2 + (y)^2 = 4$, f. $(x + 3)^2 + (y + 5)^2 = 49$, g. $(x - 5)^2 + (y - 4)^2 = 1$, h. $(x - 2)^2 + (y)^2 = 16$, i. $(x - 4)^2 + (y + 1)^2 = 9$, j. $(x + 5)^2 + (y + 5)^2 = 4$, k. $(x)^2 + (y + 3)^2 = 49$, l. $(x)^2 + (y)^2 = 4$

(30) a. $(-4, -3)$; $r = 1$, b. $(3, -3)$; $r = \sqrt{2}$, c. $(3, -5)$; $r = 7$, d. $(-3, 0)$; $r = 3$, e. $(-A, B)$; $r = \sqrt{C}$, f. $(7, 6)$; $r = \sqrt{10}$, g. $(-3, -2)$; $r = 5$, h. $(4, 3)$; $r = 10$, i. $(5, 0)$; $r = 1$, j. $(0, 0)$; $r = 7$

(31) a. II, b. III, c. I, d. IV

(32) a. B, b. C, c. D, d. C, e. C, f. C, g. A, h. D, i. D

SELF-TEST

1. Which of the following statements defines the properties of tangents?

   (a) In a plane, if a line is perpendicular to a radius of a circle at its endpoint on the circle, then the line is tangent to the circle.
   (b) If two segments from the same exterior point are tangent to a circle, then they are congruent.
   (c) If a line is tangent to a circle, then it is perpendicular to the radius drawn to the point of tangency.
   (d) All of the above

2. In the diagram, AD is a secant to the circle and AB is tangent. If the length of AC is 7 and the radius is 9, what is the length of AB?

   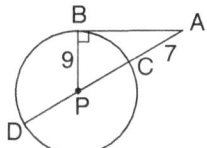

   (a) 10
   (b) 11
   (c) $5\sqrt{7}$
   (d) $7\sqrt{5}$

3. Which of the following statement defines a circle?

   (a) If a line (ℓ) is tangent to a circle, then $\ell \perp r$, where r is the radius drawn to the point of tangency.
   (b) If two circles have the same radius, then they are congruent.
   (c) A diameter is a segment that is the biggest chord in a circle.
   (d) All of the above.

4. If the length of the tangent AB is 9 cm and the secant of AD divides the length of the radius CP and AC, find the length of AC. Round your answer to the nearest tenth.

   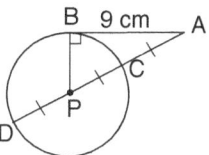

   (a) 3
   (b) 4.1
   (c) 5.2
   (d) 6.1

5. Use the diagram to find the value of x if the secants in the exterior of a circle are equal.

   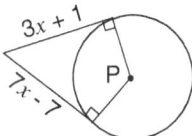

   (a) 1
   (b) 2
   (c) 3
   (d) 4

6. Which of the following statements is incorrect?

   (a) A tangent is a line in the plane of a circle that intersects the circle in exactly one point.
   (b) A chord is a segment whose endpoints are points on the circle.
   (c) A secant is a line that intersects a circle in two points.
   (d) A diameter is a chord that does not always pass through the center of the circle.

7. Which of the following is the value of r in circle P?

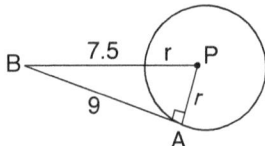

(a) 2.5
(b) 2.8
(c) 3.3
(d) 4.1

8. Which of the following is the value of r in circle P?

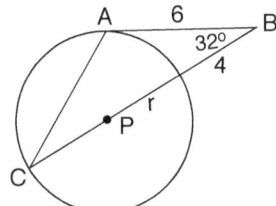

(a) 6
(b) 5
(c) 4
(d) 3

9. In the diagram, BA and BC are tangents to circle P from an external point B. The lengths of BA and BC are equal. If the radius is 4 and the length of BP is 8.5, find the length of AB.

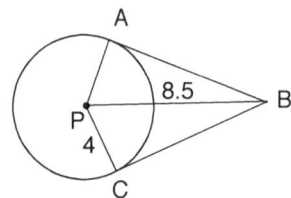

(a) 4.5
(b) 5.5
(c) 6.5
(d) 7.5

10. Which of the following statements defines the Arc Addition Postulate?

(a) The measure of an arc formed by two adjacent arcs is the sum of the measures of the two arcs.
(b) If two minor arcs of the same circle are congruent, then their central angles are congruent.
(c) If two arc lengths of two different circles are congruent, then the two arcs are congruent.
(d) All of the above.

11. What is the measure of $\widehat{AB}$?

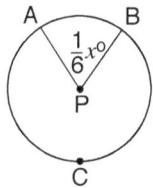

(a) 30
(b) 40
(c) 50
(d) 60

12. AP is the radius of circle and $m\angle APB$ is 90°. Find the value of x.

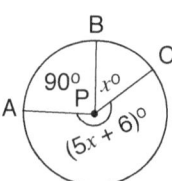

(a) 30
(b) 38
(c) 44
(d) 48

13. In the diagram, the length of BE equals the length of AE in circle P. If the length of BE is 9 units and the radius of circle is 11 units, what is the length of PE?

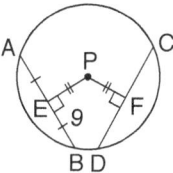

(a) $2\sqrt{10}$
(b) $3\sqrt{10}$
(c) 5
(d) 6

14. Which of the following statements does not define an arc?

(a) If the arc measure in two circles is congruent, then their arc lengths are congruent.
(b) If the radii of two circles are congruent, then their arc lengths are congruent.
(c) If two arc lengths of the same circle are congruent, then two arcs are congruent.
(d) All of the above.

15. AB is the diameter of the circle below. Find the value of x.

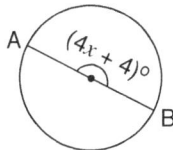

(a) 40
(b) 42
(c) 44
(d) 46

16. In the diagram, AB and CD are congruent in circle P. If the length of AB is 24 cm and the length of PE is 9 cm, what is the length of the radius in circle P?

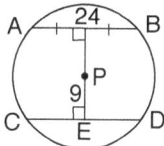

(a) 7
(b) 10
(c) $4\sqrt{7}$
(d) $3\sqrt{7}$

17. Find the value of x.

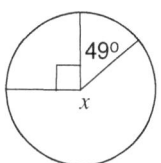

(a) 211
(b) 221
(c) 231
(d) 241

18. Which of the following statements defines the chords of circles?

(a) In congruent circles, two minor arcs are congruent if and only if their corresponding chords are not congruent.
(b) If a diameter of a circle is perpendicular to a chord, then the diameter does not bisects the chord and its arc.
(c) If one chord is a perpendicular bisector of another chord, then the first chord is a diameter.
(d) In congruent circles, two chords are congruent if and only if they are not equidistant from the center

19. What is the value of x?

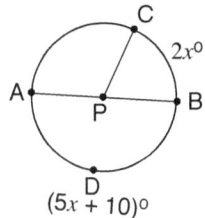

(a) 30
(b) 32
(c) 34
(d) 36

20. $\widehat{AB}$, $\widehat{BC}$, and $\widehat{AC}$ are congruent. What is the measure of ∠BPC?

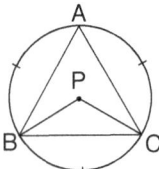

(a) 110°
(b) 120°
(c) 130°
(d) 140°

21. Which of the following statement defines inscribed polygons?

(a) If two inscribed angles of a circle intercept the same arc, then the angles are congruent.
(b) If an angle is inscribed in a circle, then its measure is half the measure of its intercepted arc.
(c) A quadrilateral can be inscribed in a circle if and only if its opposite interior angles are supplementary.
(d) All of the above.

22. $m\widehat{AB}$ is 109° and $m\widehat{AC}$ is 79°, find the measure of the value of x.

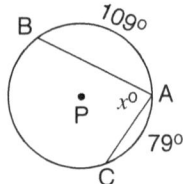

(a) 92°
(b) 90°
(c) 88°
(d) 86°

23. If $m\angle BAC$ is 35°, find the measure of $\widehat{BC}$.

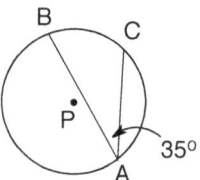

(a) 35°
(b) 45°
(c) 70°
(d) 90°

24. The measure of $\widehat{AC}$ is 60°. Find the measure of the value of x.

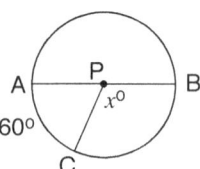

(a) 120°
(b) 110°
(c) 100°
(d) 130°

25. Which of the following statement defines the properties of an inscribed polygon?

(a) If $m\angle A = 92°$ in a quadrilateral is inscribed in a circle, then the opposite angle is 92°.
(b) If the right angle is inscribed in a circle, then the measure of the inscribed arc is 90°.
(c) A quadrilateral can be inscribed in a circle if and only if the sum of its opposite angles is 90°.
(d) If a right triangle is inscribed in a circle, then its hypotenuse is a diameter of the circle.

26. The measure of $\widehat{AC}$ is 70°. Find the measure of the value of x.

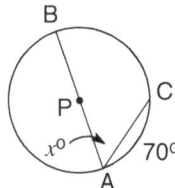

(a) 55°
(b) 60°
(c) 120°
(d) 110°

27. Find the value of x.

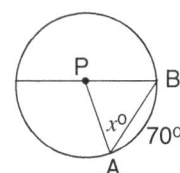

(a) 55°
(b) 60°
(c) 65°
(d) 70°

28. The measure of $\widehat{AC}$ is 128° and the measure of $\widehat{AB}$ is 108°, find the value of x.

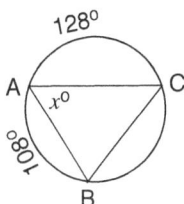

(a) 62°
(b) 60°
(c) 68°
(d) 66°

29. The length of $m\widehat{AB}$ is 170° and $m\widehat{AC}$ is 160°. Find the measure of $\angle BAC$.

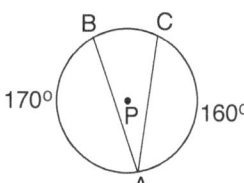

(a) 10°
(b) 15°
(c) 20°
(d) 25°

30. $\angle B$ and $\angle D$ are inscribed angles of a circle that intercept the same arc. If $m\angle D$ is 27°, find the value of x.

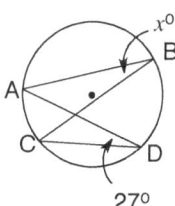

(a) 54°
(b) 27°
(c) 30°
(d) 24°

31. If $m\angle BPD$ is 52°, which of the following is the greatest angle?

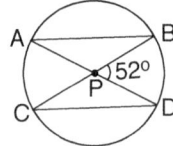

(a) $m\angle ABC$
(b) $m\angle APB$
(c) $m\angle BCD$
(d) $m\angle APC$

32. If $m\widehat{AC}$ is 125°, find the value of x.

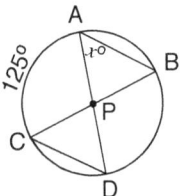

(a) 62.5°
(b) 65°
(c) 55°
(d) 63.5°

33. $\angle B$ and $\angle C$ are inscribed angles of a circle that intercept the same arc. If $m\angle C$ is 62°, find the value of x.

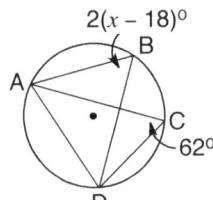

(a) 47
(b) 48
(c) 49
(d) 50

34. The measure of $\angle ACB$ is 48°, find the measure of the value of x.

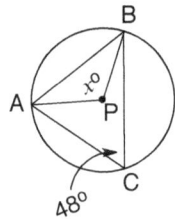

(a) 48°
(b) 96°
(c) 90°
(d) 98°

35. If the measure of $\angle APD$ is 94°, find the value of x.

(a) $15\frac{1}{3}$
(b) 16
(c) $16\frac{1}{3}$
(d) 15

36. Find $m\widehat{AD}$.

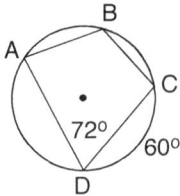

(a) 48°
(b) 96°
(c) 146°
(d) 156°

37. Which of the following statements defines an angle relationship in the given diagram below?

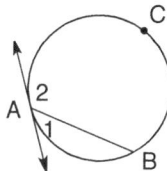

(a) $m\angle 1 = \frac{1}{2} m\widehat{AB}$
(b) $m\angle 2 = \frac{1}{2} m\widehat{ACB}$
(c) If $m\angle 2 = 156°$, then $m\widehat{ACB} = 78°$.
(d) If $m\widehat{AB} = 78°$, then $m\widehat{ACB} = 282°$.

38. Which of the following statements is not a definition of an angle relationship in circles given the diagram?

(a) If two tangents intersect in the interior of a circle, the measure of the angle formed is one half the difference of the measures of the intercepted arcs.
(b) If a tangent intersects in the exterior of a circle, the measure of the angle formed is one half the difference of the measures of the intercepted arcs.
(c) If two chords intersect in the interior of a circle, then the measure of each angle is one half the sum of the measures of the arcs intercepted by the angle and its vertical angle.
(d) If two secants intersect in the interior of a circle, then the measure of each angle is one half the sum of the measures of the arcs intercepted by the angle and its vertical angle.

39. What is the value of x?

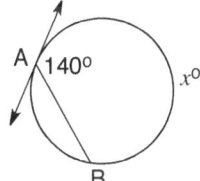

(a) 120°
(b) 140°
(c) 280°
(d) 240°

40. The measure of $\widehat{BC}$ is 94°. Find the value of x.

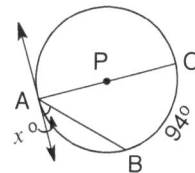

(a) 86°
(b) 43°
(c) 94°
(d) 47°

41. $\widehat{ADC}$ is a semicircle. Find the value of x.

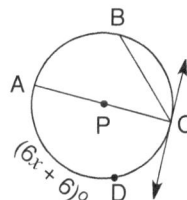

(a) 55
(b) 57
(c) 59
(d) 61

42. Two chords AD and BC intersect in the interior of a circle. Find the value of x.

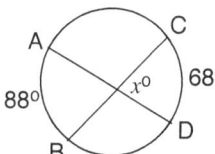

(a) 68°
(b) 78°
(c) 88°
(d) 98°

102   Chapter 9   Circles

43. In the diagram below, $m\widehat{BC} = 100°$ and $m\angle BAC = 37°$. Find the value of x.

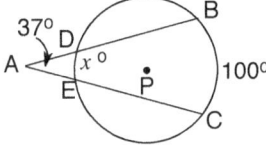

(a) 43°
(b) 37°
(c) 27°
(d) 26°

44. Tangent AB and secant AC intersect in the exterior of circle P. If $m\widehat{BD} = 178°$ and $m\angle BAC = 60°$, what is the measure of the intercepted arc?

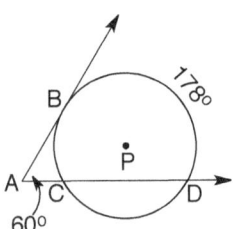

(a) 48°
(b) 58°
(c) 68°
(d) 89°

45. In the diagram, find the measure of $\widehat{BD}$ given $m\angle AFE = 24°$ and $m\angle BAF = 30°$.

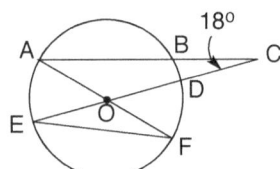

(a) 12°
(b) 11°
(c) 10°
(d) 9°

46. In the diagram, find the measure of $\angle C$ if $m\angle FED = 22°$ and $m\angle BAF = 30°$.

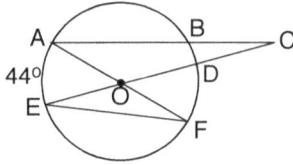

(a) 8°
(b) 14°
(c) 18°
(d) 20°

47. Two chords AD and BC intersect in the interior of circle P. Given that $m\widehat{CD} = 64°$, find the measure of the value of $x°$.

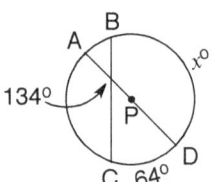

(a) 116°
(b) 124°
(c) 134°
(d) 152°

48. Chords AD and BC intersect in the interior of a circle. $m\widehat{AC} = 112°$ and $m\widehat{BD} = 94°$ Find the value of x.

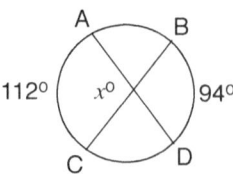

(a) 98°
(b) 103°
(c) 112°
(d) 94°

49. AB is tangent to circle P. Arc measure BC is 128°. Find the measure of ∠A.

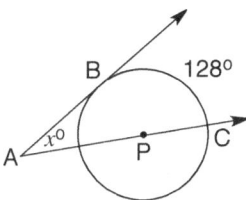

(a) 22°
(b) 38°
(c) 52°
(d) 76°

50. Find the value of x.

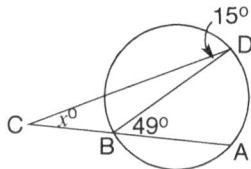

(a) 30°
(b) 68°
(c) 34°
(d) 65°

51. Find the measure of ∠A. If necessary, round your answer to the nearest tenth.

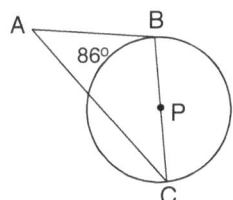

(a) 24°
(b) 68°
(c) 74°
(d) 94°

52. In the diagram, the measures of $\widehat{AB}$, $\widehat{AC}$, and $\widehat{CD}$ are congruent. What is the measure of ∠E if the measure of $\widehat{ACD}$ is 210°?

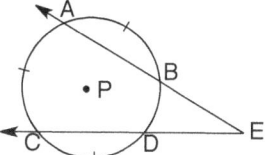

(a) 30°
(b) 40°
(c) 50°
(d) 60°

53. Which of the following statement defines the tangent and secant in the diagram below?

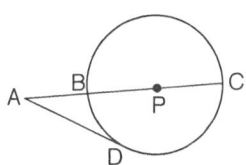 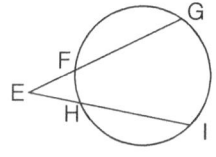

(a) The length of AD is a tangent segment.
(b) The length of AC, EG, or EI is a secant segment.
(c) The length of AB is an external secant segment.
(d) All of the above.

54. Given the diagram from Question 53, which of the following statements defines the segment of tangent and secant?

(a) $(AD)^2 = (AB)(AC)$
(b) $(EF)(EG) = (EH)(EI)$
(c) If the length of AB is the same length of the radius, then $\overline{AD} = r\sqrt{3}$.
(d) All of the above.

55. Two chords intersect in the interior of a circle. Find the value of x.

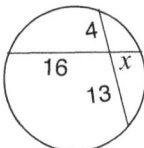

(a) 3.12
(b) 3.25
(c) 3.50
(d) 3.75

56. Two secant segments AE and AC share the same endpoint outside a circle. The length of AD equals the length of DE. AB = 8 and BC = 11.5. Find the length of AD. Round your answer to the nearest tenth.

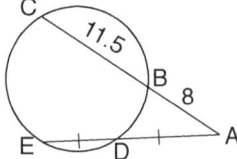

(a) 8.0
(b) 8.4
(c) 8.8
(d) 9.2

57. AC is a secant segment and AD is a tangent segment that shares the same endpoint A outside of the circle. Find the value of x.

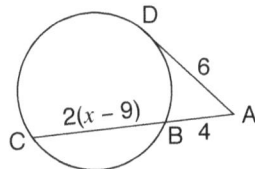

(a) 9.5
(b) 10.5
(c) 11.5
(d) 12.5

58. AB is a tangent in circle P and ACD is a secant. Find the length of BD. If necessary, round your answer to the nearest tenth.

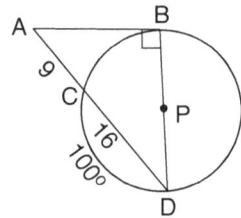

(a) 22
(b) 20
(c) 18
(d) 16

59. What is the standard equation of a circle with a diameter of 10 and its center at P(4, −2)?

(a) $(x + 4)^2 + (y - 2)^2 = 5$
(b) $(x + 4)^2 + (y - 2)^2 = 100$
(c) $(x - 4)^2 + (y + 2)^2 - 25 = 0$
(d) $(x - 4)^2 + (y - 2)^2 - 10 = 0$

60. What is the standard equation of a circle with its center at P(0, 0) and a radius of 2?

(a) $(x)^2 + (y)^2 = 2$
(b) $(x)^2 + (y)^2 = 0$
(c) $(x)^2 + (y)^2 - 2 = 0$
(d) $(x)^2 + (y)^2 - 4 = 0$

61. Find the coordinates for the center and radius of the circle by using the standard equation of $(x - 5)^2 + (y + 9)^2 = 5$.

(a) $c(5, -9)$ and $r = 5$
(b) $c(-5, 9)$ and $r = 5$
(c) $c(5, -9)$ and $r = \sqrt{5}$
(d) $c(5, -9)$ and $r = 25$

62. Find the quadrant of the coordinates of the center of a circle with the given equation of $(x + 5)^2 + (y - 7)^2 = 2$.

(a) I
(b) II
(c) III
(d) IV

63. Which of the following is the equation of a circle with a radius of $\sqrt{2}$ and its center at P(4, −2)?

(a) $(x - 4)^2 + (y + 2)^2 = 2$
(b) $(x + 4)^2 + (y - 2)^2 = \sqrt{2}$
(c) $(x + 4)^2 + (y - 2)^2 - \sqrt{2} = 0$
(d) $(x - 2)^2 + (y + 4)^2 - 4 = 0$

64. Which of the following is the equation of a circle with a diameter of 100 and its center at P(2, −2)?

(a) $(x - 2)^2 + (y + 2)^2 = 10$
(b) $(x - 2)^2 + (y + 2)^2 = 100$
(c) $(x - 2)^2 + (y + 2)^2 = 5$
(d) $(x - 2)^2 + (y + 2)^2 = 25$

65. Which of the following is the equation of a circle with a radius of 9 and its center at (0, 0)?

(a) $(x)^2 + (y)^2 - 1 = 9$
(b) $(x)^2 + (y)^2 + 1 = 10$
(c) $(x)^2 + (y)^2 - 1 = 81$
(d) $(x)^2 + (y)^2 + 1 = 82$

66. What is the radius of the circle with equation $(x)^2 + (y)^2 - 8y = -12$?

(a) 2
(b) 4
(c) $2\sqrt{3}$
(d) 12

ANSWERS

(1) d	(2) c	(3) d	(4) c	(5) b	(6) d
(7) c	(8) b	(9) d	(10) d	(11) d	(12) c
(13) a	(14) d	(15) c	(16) d	(17) b	(18) c
(19) c	(20) b	(21) d	(22) d	(23) c	(24) a
(25) d	(26) a	(27) a	(28) a	(29) b	(30) b
(31) b	(32) a	(33) c	(34) b	(35) a	(36) d
(37) c	(38) a	(39) c	(40) b	(41) c	(42) b
(43) d	(44) b	(45) a	(46) b	(47) d	(48) b
(49) b	(50) c	(51) c	(52) a	(53) d	(54) d
(55) b	(56) c	(57) c	(58) b	(59) c	(60) d
(61) c	(62) b	(63) a	(64) d	(65) d	(66) a

CHAPTER 10
Area of Polygons and Circles

In this chapter, you will find the angle measures in polygons, the areas of regular polygons as well as the perimeters and areas of similar figures, the circumference and arc length of circles, and the areas of circles and their sectors.

[CONCEPTS] [EXAMPLES] [FORMULAS] [VOCABULARY]

10-1. Formulas for finding the angle measures, areas, or circumference of a polygon.

Name of Polygon	Formula (Angles, Areas, or Circumference)	
Polygon Interior Angles	n-gon = $(n-2)(180°)$	n = number of side
Polygon Exterior Angles	n-gon = $\frac{1}{n}(360°)$	n = number of side
Area of an Equilateral Triangles	$A = \frac{1}{4}\sqrt{3}(s^2)$	s = side length
Area of a Regular Polygon	$A = \frac{1}{2}aP$ or $A = \frac{1}{2}a(ns)$	a = apothem, P = perimeter, s = side length, and n = number of side
Circumference of a Circle	$C = \pi d$ or $C = 2\pi r$	r = radius and d = diameter
Area of a Circle	$A = \pi r^2$	r = radius
Area of a Sector	$\frac{\text{Arc length of } \widehat{AB}}{2\pi r} = \frac{m\widehat{AB}}{360°}$	r = radius, $m\widehat{AB}$ = measure of arc length

10-2. Find the sum of the measures of the interior angles of the convex n-gon.

   a. 10-gon b. 4-gon

> **SOLUTION**
>
> By the definition of polygon interior angles, show that the sum of the measures of the interior angles of a convex n-gon is $(n-2)(180°)$, where n is the number of the sides.
> a. Sum of 10-gon = $(n-2)(180°) = (10-2)(180°) = 1440°$
> b. Sum of 4-gon = $(n-2)(180°) = (4-2)(180°) = 360°$

10-3. Find the measure of the remaining interior angle of a convex n-gon.
   6-gon; 94°, 92°, 94°, 126°, 152°, and _____

Chapter 10   Area of Polygons and Circles

SOLUTION

First, find the sum of the measures of the interior angles of a 6-gon. Then find the sum of the measures of the given interior angles. You will then be able find the measure of the unknown interior angle by subtracting the sum of the measures of the interior angles from the sum of the 6-gon's interior angles.
Interior angles measures of a 6-gon = $(n-2)(180°) = (6-2)(180°) = 720°$
Sum of the given angles = $(94° + 92° + 94° + 126° + 152°) = 558°$
$x = 720° - 558° = 162°$

10-4. Find the value of x.

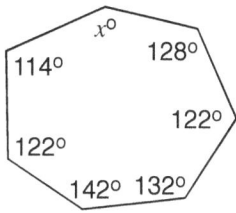

SOLUTION

First, find the sum of the measures of the interior angles of a 7-gon. Then find the sum of the measures of the given interior angles. To find the unknown angle measure, insert the measures and the value of x in an equation and solve for x.

Sum of the n-gon's interior angles = $(n-2)(180°)$, where n represents the number of the sides.
7-gon Interior Angles Measure = $(7-2)(180°) = 900°$
Sum of the given angles = $(114° + 122° + 142° + 132° + 122° + 128° + x) = 900°$
$(760° + x) = 900°$
$x = 140°$

Therefore, the measure of the remaining interior angle x is $140°$.

10-5. Find the value of x.

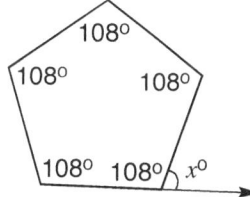

SOLUTION

By the definition of polygon exterior angles, the sum of the measures of the exterior angles of a convex polygon is 360°. So the measure of each exterior angle of a regular n-gon is $\frac{1}{n}(360°)$, where n is the number of the sides.

Sum of the exterior n-gon = $(1/n)(360°)$
$$x = (1/5)(360°)$$
$$x = 72°$$

Therefore, the measure of the remaining exterior angle x is 72°.

10-6. Find the area of the equilateral triangle.

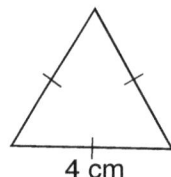

4 cm

SOLUTION

By the definition of the area of an equilateral triangle, you can show that the area of an equilateral triangle is four times the product of the square of the length of the side and $\sqrt{3}$. So the <u>area of an equilateral triangle</u> is $\frac{1}{4}\sqrt{3}s^2$, where s represents the length of the side.

$A = \frac{1}{4}\sqrt{3}s^2$ Use the formula for the area of an equilateral triangle.

$A = \frac{1}{4}\sqrt{3}(4)^2$ Substitute.

$= 4\sqrt{3}$ cm^2 Simplify.

The area of the equilateral triangle is $4\sqrt{3}$ cm^2.

10-7. Find the perimeter and area of the square below. If necessary, round your answer to the nearest tenth.

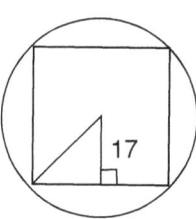

17

SOLUTION

By the definition of the area of a regular polygon, the area of a regular n-gon with side length (s) is half the product of the apothem (a) and the perimeter (P).

Chapter 10 Area of Polygons and Circles

The <u>apothem</u> of the polygon is the distance from the center of a perpendicular line to any side of the polygon like the figure below.

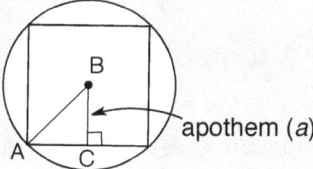

The perimeter of a square is the sum of the lengths of the four sides. $P = n(S) = 4(34) = 136$ units. The area of a square is $A = (1/2)aP$ or $A = (1/2)a(nS)$.
$A = (1/2)a(nS) = (1/2)17(4)(34) = 1156$ units^2
So the perimeter of a square is 136 units and the area of a square is 1156 units^2.

10-8. Find the shaded and unshaded areas of the regular polygon below. If necessary, round your answer to the nearest tenth.

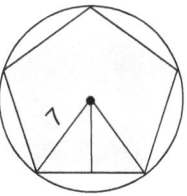

SOLUTION

You can first find the measure of the central angle of the pentagon. The central angle of a regular pentagon is 360°/5 or 72°. So the top angle in the apothem is 36° because the apothem bisects it. Now you can find the apothem by using trigonometry.

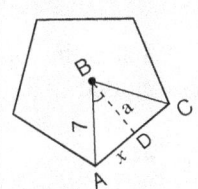

$\cos(36°) = \dfrac{a}{7}$

$a = 7\cos(36°)$

You can also find half of the side length ($\overline{AD}$) of the pentagon.
$\sin(36°) = \dfrac{x}{7}$, where x is the length of AD.

$x = 7\sin(36°)$
So AC = $2x = (2)(7\sin(36°))$
$ = 14\sin(36°)$

So the length of AC of the pentagon is 14 sin(36°). Since you now know a side length of the pentagon, you can find the perimeter. $P = 5(14\sin(36°)) = 70\sin(36°)$. Finally, you can find the area of the pentagon.

$A = \dfrac{1}{2}aP$ \qquad Use the area of a regular polygon.

$ = \dfrac{1}{2}7\cos(36°)70\sin(36°)$ \qquad Substitute.

≈ 116.5 units²   Find by using a calculator
To find the area of a circle.
$A = \pi r^2$.   Use the area of a circle.
   $= \pi (7^2)$ Substitute.
   ≈ 153.9 units²

The area of the unshaded region can be found by subtracting the area of the polygon from the area of a circle.
$A_{unshaded} = 153.9 - 116.5 = 37.4$ units²
So the shaded area is about 116.5 units² and the unshaded area is about 37.4 units².

* Further information for finding an apothem in each polygon.
First find the top angle of the triangle in each polygon. The central angle of any polygon is then found by dividing the number of sides of an *n*-gon by 360°. For instance, the central angle of a pentagon is 360°/5 =72°. Afterwards, divide the central angle of the triangle to find the measure of ∠ABD or ∠CBD. Now you are able to find an apothem (*a*) of each polygon for using the information below.

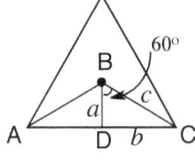
For a 30°-60°-90° triangle
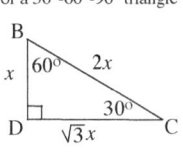

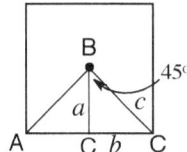
For a 45°-45°-90° triangle
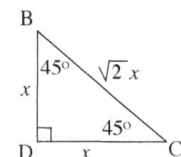

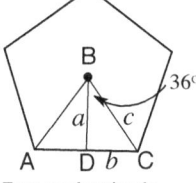
For a regular triangle
$a = (\cos 36°)(c)$
$b = (\sin 36°)(c)$
$b = (\tan 36°)(a)$
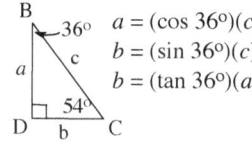

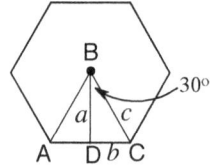
For a 30°-60°-90° triangle
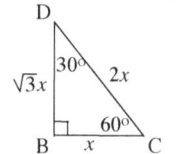

10-9. The ratio of the two similar rectangles is shown below. What is the ratio of the side lengths? What is the ratio of the areas of the hexagons A and B?

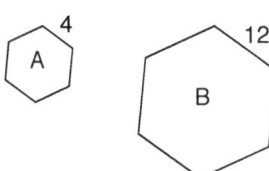

SOLUTION

By the definition of the areas of similar polygons, if two polygons are similar with the lengths of corresponding sides being the ratio of $a : b$, then the ratio of their areas is $a^2 : b^2$. The ratio of the lengths of the corresponding sides in the pictures A and B is 4 : 12 or 1 : 3. The ratio of their area is $4^2 : 12^2$ or 1 : 9.

10-10. Two similar triangles have areas of 12 in.² and 108 in.². What is the ratio for the lengths of the corresponding sides?

> **SOLUTION**
>
> The ratio of the areas is 12 : 108 or 1 : 9. So the ratio of the corresponding sides is 1: 3 by the definition of the area of similar polygons that are the lengths of corresponding sides being the ratio of $a : b$ and the ratio of their areas is $a^2 : b^2$.

10-11. The ratio of the two similar rectangles is shown below. The areas of the rectangle A and B are 2 feet by 4 feet and 12 feet by 24 feet respectively. What is the ratio of the side lengths? What is the ratio of the areas?

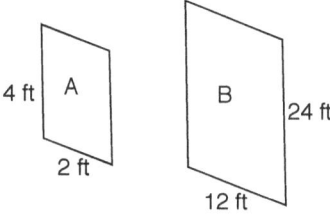

> **SOLUTION**
>
> The ratio for the lengths of the corresponding sides in the rectangles A and B is 2 : 12 or 1 : 6. The ratio of the areas is $1^2 : 6^2$ or 1 : 36 by the definition of the areas of similar polygons.

10-12. Find the circumference of a circle.

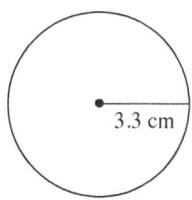

> **SOLUTION**
>
> By the definition of the circumference (C) of a circle, C = πd or C = 2πr, where d is the diameter of the circle and r is the radius of the circle.
> C = 2πr   Use the formula for the circumference of a circle.
> C = 2π(3.3) = 6.6π cm.   Substitute and simplify.
> Therefore, the circumference of a circle is 6.6π cm.

10-13. Find the length of $\widehat{AB}$.

a.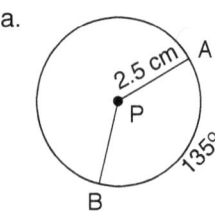

b.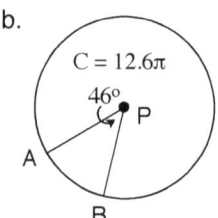

SOLUTION

a. By the Arc Length Corollary, in a circle, the ratio of the length of a given arc to the circumference is equal to the ratio of the arc measure to 360°.

$$\frac{\text{Arc length of } \widehat{AB}}{2\pi r} = \frac{m\widehat{AB}}{360°} \text{ or Arc length of } \widehat{AB} = \frac{m\widehat{AB}(2\pi r)}{360°}$$

So, Arc length of $\widehat{AB} = \frac{m\widehat{AB}(2\pi r)}{360°}$

$= \frac{135°(2\pi 2.5)}{360°}$

$= 1.875\pi$ cm

b. Arc length of $\widehat{AB} = \frac{m\widehat{AB}(C)}{360°}$

$= \frac{46°(12.6\pi)}{360°}$

$= 1.61\pi$ unit

So the arc length of AB is 1.61π units.

10-14. Find the area of the circle. Leave your answer in terms of π.

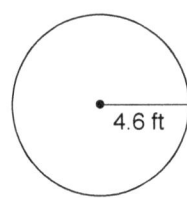

SOLUTION

By the definition of an area of a circle, the area of a circle is π times the square of the radius, or $A = \pi r^2$.

So, $A = \pi r^2$ Use the formula of the area of a circle.

$= \pi(4.6)^2$ Substitute.

$= 21.16\pi$ ft^2 Simplify.

Therefore, the area of a circle is 21.16π ft^2.

Chapter 10   Area of Polygons and Circles

10-15. Find the indicated dot area of the sector of a circle. Round to the nearest hundredth.

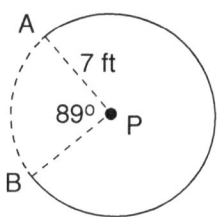

SOLUTION

By the definition of the area of the sector of a circle, the ratio of the area (A) of a sector of a circle to the area of the circle is equal to the ratio of the intercepted arc measure to 360°.

$$\frac{\text{Area A}}{\pi r^2} = \frac{m\widehat{AB}}{360°}, \text{ or Area A} = \frac{m\widehat{AB}(\pi r^2)}{360°}$$

So, Area A $= \frac{m\widehat{AB}(\pi r^2)}{360°}$

$= \frac{89°(\pi 7^2)}{360°}$

$\approx 38.04 \text{ ft}^2$

Therefore, the area of a sector of a circle is 38.04 ft².

10-16. Find the area of the circle and the area of the shaded region.

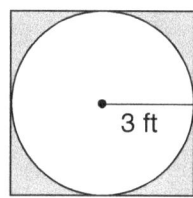

SOLUTION

First, find the area of the circle.
To find the area of the circle:
The area of a circle is π times the square of the radius, or $A = \pi r^2$.
So, $A = \pi r^2$ Use the area of a circle.
$= \pi(3)^2$ Substitute.
$= 28.26 \text{ ft}^2$ Simplify.
To find the area of the square:
By the definition of the area of a square, the area of a regular n-gon with side length S is half the product of the apothem (a) and the perimeter P.
$A = (1/2)aP$ or $A = (1/2)a(nS)$
The apothem is 3 ft. The side length of the regular square is twice the length of the apothem a. So, S = 6 ft.
$A = (1/2)a(nS) = (1/2)3(4)(6) = 36 \text{ ft}^2$
You can now find the area of the shaded region by subtracting the area of a circle from the area of the square.
Area of shaded region = 36 − 28.26 = 7.74 ft²

PRACTICES

1. Answer each question.
《See Example 10-2》

   a. What is a convex polygon?

   b. What is a concave polygon?

   c. What is a regular polygon?

   d. What is the formula of the sum of interior angles for a convex n-gon?

   e. What is the formula of each exterior angle of a regular n-gon?

2. Find the sum of the interior angles for each polygon.
《See Example 10-2》

   a. Heptagon   b. 15-gon

   c. 25-gon   d. Square

   e. 12-gon   f. 36-gon

   g. Triangle   h. Pentagon

3. Find the measure of the unknown interior angle of each convex n-gon.
《See Example 10-3》

   a. 6-gon; 130°, 85°, 96°, 138°, 143°, and _____

   b. 5-gon; 86°, 109°, 91°, 114°, and _____

   c. 6-gon; 97°, 122°, 128°, 98°, 136°, and _____

   d. 8-gon; 138°, 125°, 117°, 135°, 131°, 152°, 97°, and _____

   e. 10-gon; 182°, 145°, 113°, 137°, 142°, 148°, 136°, 162°, 132°, and _____

4. Find the value of x.
《See Example 10-4》

a.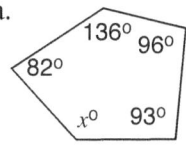

b.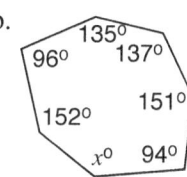

c.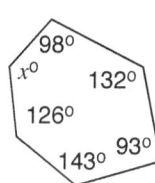

5. Find the measure of the exterior angle of each regular n-gon.
《See Example 10-2》

    a. Hexagon   b. 13-gon

    c. 12-gon   d. 24-gon

    e. Octagon   f. 10-gon

6. Find the measure of one exterior angle in each polygon. If necessary, round your answer to the nearest whole number.
《See Example 10-2》

a.

b.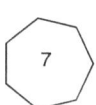

c.

7. Find the sum of the interior angles of each convex n-gon.
《See Example 10-2》

    a. Find the sum of the interior angles of a regular 14-gon.

    b. What is the measure of an interior angle of a regular 12-gon?

    c. What is the sum of the measures of the interior angles of a regular 18-gon?

    d. What is the measure of an exterior angle of a regular 11-gon?

    e. What is the sum of the measures of the exterior angles of a convex octagon?

116 Chapter 10 Area of Polygons and Circles

8. Find the number of sides of each described polygon.
《See Example 10-2》

   a. The measure of each interior angle of a regular polygon is 156°. How many sides does the polygon have?

   b. The measure of each interior angle of a regular polygon is 172.5°. How many sides does the polygon have?

   c. The sum of the interior angle measures of a convex polygon is 3060°. How many sides does the polygon have?

   d. The sum of the interior angle measures of a convex polygon is 2160°. How many sides does the polygon have?

9. Find the value of x.
《See Example 10-5》

   a. 117° 99° 78° x°

   b. 82° x° 102° 125° 93°

   c. 82° (5x - 22)° 125° 93° (3x + 6)°

   d. 115° (2x +12)° 4x° 155°

   e. (5x - 5)° 80° (2x + 10)° 4x°

   f. (6x - 10)° (3x - 4)° 2x°

10. Find the area of each equilateral triangle given. Round to the nearest hundredth.
《See Example 10-6》

   a. 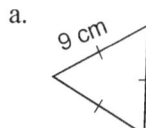 9 cm

   b. 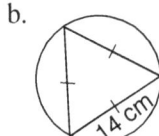 14 cm

   c.  5 cm

11. Answer each problem about the polygon below.
≪See Examples 10-7 and 10-8≫

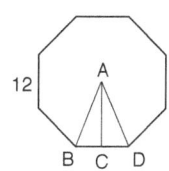

a. Name the polygon.

b. Find the length of CD.

c. Name the apothem of the polygon.

d. Find the length of BD.

e. Find the length of AD.

f. Find the perimeter of the polygon.

g. Find the measure of the central angle.

h. Find the area of the polygon.

i. Find the measure of ∠CAD.

12. Find the perimeter and area of the regular polygon. If necessary, round to the nearest tenth.
≪See Examples 10-7 and 10-8≫

a. 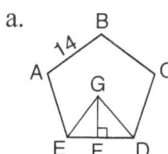 b. 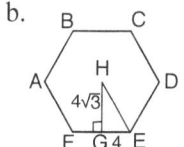 c.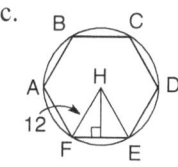

d. 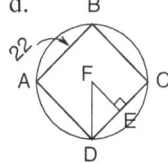 e. 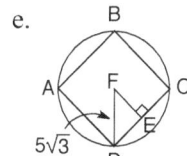 f.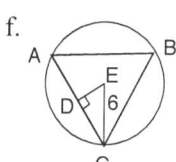

13. Find the perimeter and area of each polygon described. If necessary, round to the nearest tenth.
    ≪See Examples 10-7 and 10-8≫

    a.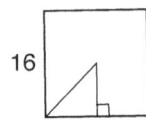
    b.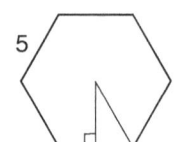
    c.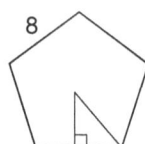

    d.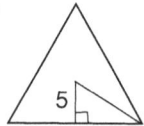
    e.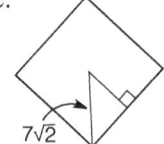
    f.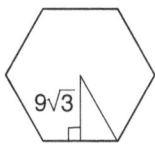

14. Find the area of each given description.
    ≪See Examples 10-7 and 10-8≫

    a. hexagon
       length of apothem = 15.6
       length of a side = 18

    b. pentagon
       length of apothem = 19.5
       length of a side = 28

    c. hexagon
       length of apothem = 8.7
       length of a side = 9

    d. heptagon
       length of apothem = 15
       length of a side = 14.4

15. Find the shaded and unshaded areas of the regular polygon. If necessary, round your answer to the nearest tenth.
    ≪See Examples 10-1 and 10-12≫

    a.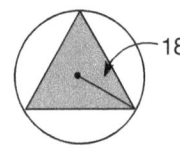
    b.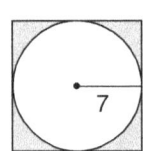
    c.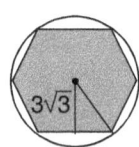

    d.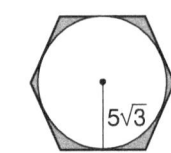
    e.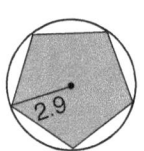
    f.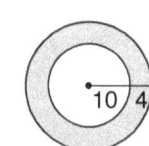

16. The ratio of the two similar rectangles is shown below. What is the ratio of the side lengths? What is the ratio of the areas of hexagons A and B?

≪See Example 10-9≫

a. 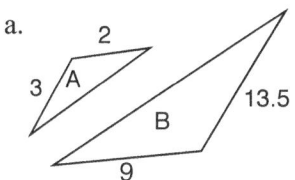   b. 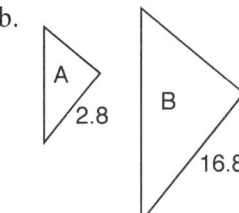   c.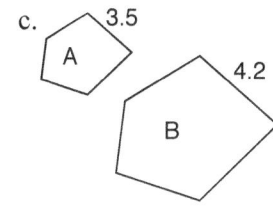

17. Find the solution for the problems below.

≪See Example 10-10≫

   a. There are two similar pentagons with areas of 24 cm² and 156 cm². Find the similarity ratio of the pentagons.

   b. If the ratio of the areas of two similar triangles is 6 : 4, find the ratio of the lengths of corresponding sides.

   c. If two similar triangles have areas of 8 in.² and 112 in.², find the similarity ratio of each pair of triangles.

   d. If the ratio of the lengths of corresponding sides in the squares is 1 : 7, find the ratio of the perimeters.

   e. If the ratio of the lengths of corresponding sides in the regular pentagon is 5 : 25, find the ratio of the areas.

   f. The areas of two similar trapezoids are 4 ft² and 32 ft². What is each similarity ratio of the areas?

   g. The areas of two similar circles are 9π cm² and 49.5π cm². What is the similarity ratio of the areas and perimeters?

   h. If the ratio of two similar trapezoids is 1 : 3, find the ratio of the areas.

18. For Exercises **a-c**, find the ratio of the perimeters of the similar triangles using the two given areas.
≪See Examples 10-10 and 10-11≫

a. A = 9.61π units²
   B = 29.16π units²

b. 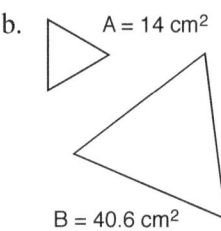 A = 14 cm²
   B = 40.6 cm²

c. 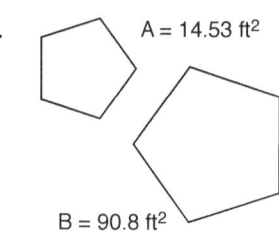 A = 14.53 ft²
   B = 90.8 ft²

19. Use the figure shown below for Exercises **a-d**. It is given that △ABC ~ △ADE. Round your answer to the nearest tenth.
≪See Examples 10-10 and 10-11≫

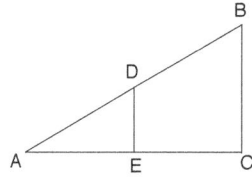

a. If the two similar triangles had areas of 10 cm² and 40 cm², find the similarity ratio of each pair of triangles.

b. AD = 5 ft and AB = 20 ft and the area of △ADE is 24 ft². Find the area of △ABC.

c. AD = 12 cm and AB = 62.4 cm and the area of △ADE is 48 cm². Find the area of △ABC.

d. The ratio of the areas of △ABC and △ADE is 24 : 3. Find the similarity ratio of △ABC and △ADE.

20. Use the diagram of the circle shown. Find the following lengths in each problem. If necessary, round to the nearest tenth.
≪See Example 10-13≫

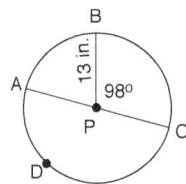

a. Find $m\widehat{AB}$.

b. Find $m\widehat{BC}$.

c. Find the circumference of the circle.

d. Find the length of $\widehat{AB}$.

e. Find the length of $\widehat{BC}$.

f. Find the length of $\widehat{ADC}$.

g. Find the area of the circle.

h. Find $m\angle APB$.

21. Find the circumference and the length of each arc.
《See Examples 10-12 and 10-13》

   a. Find the circumference of the circle.

   b. Find the length of $\widehat{AB}$.

   c. Find the length of $\widehat{BC}$.

   d. Find the length of $\widehat{CD}$.

   e. Find the length of $\widehat{DE}$.

   f. Find the length of $\widehat{AED}$.

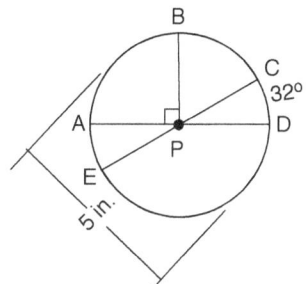

22. Find the circumference of each circle. Round to the nearest hundredth if necessary.
《See Example 10-12》

   a.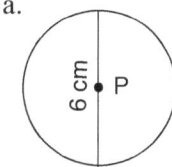

   b.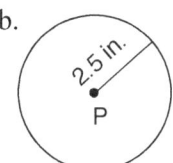

   c.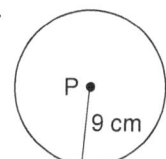

   d.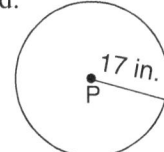

   e.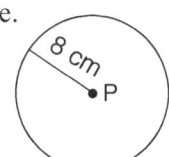

   f.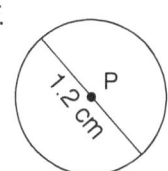

23. Find the length of $\widehat{AB}$ of the circle. Round to the nearest tenth.
《See Example 10-13》

   a.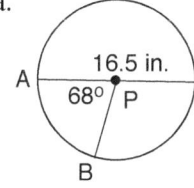

   b.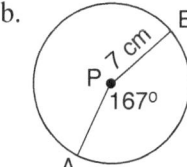

   c.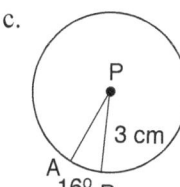

d. 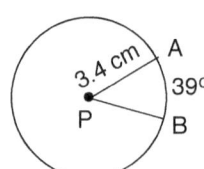 e. 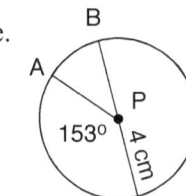 f.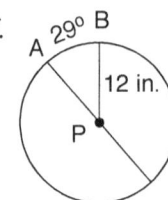

24. Solve the following problems.
≪See Example 10-13≫

a. Find the radius of a circle that has a circumference of 10.4π cm.

b. Find the diameter of a circle that has a circumference of 20π in.

c. Find the circumference that has a radius of 8 in.

d. Find the circumference that has a diameter of 14 cm.

e. Find the arc length of a circle that has a radius of 6 ft and an angle measure of 65°.

f. Find the arc length of a circle that has a diameter of 10 cm and an angle measure of 23°.

25. Find the following values.
≪See Example 10-13≫

a. circumference of circle P   b. circumference of circle P   c. length of $\widehat{AB}$

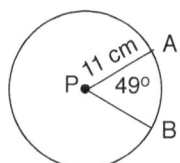   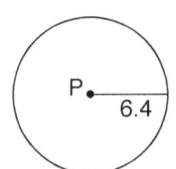   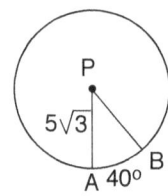

d. length of $\widehat{AB}$ e. radius of circle P   f. i) circumference of inner circle P
                                                          ii) circumference of outer circle P

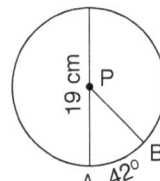   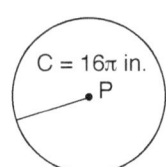   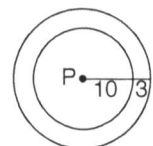

26. Find the measure of the indicated dot area of the circle. Round to the nearest tenth.
≪See Example 10-15≫

a.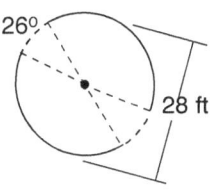

b.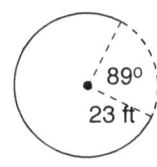

c.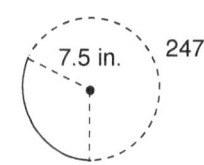

d.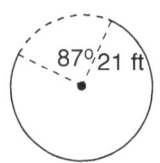

e.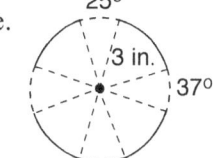

f.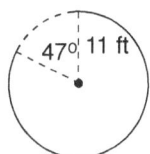

g.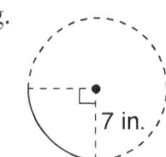

h.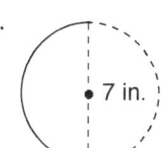

i.

27. Find the following values for each circle below. Round to the nearest tenth.
≪See Examples 10-13 and 10-14≫

a. length of $\widehat{AB}$

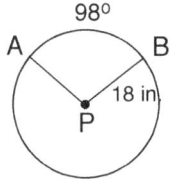

b. length of $\widehat{AB}$

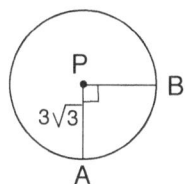

c. circumference of circle P

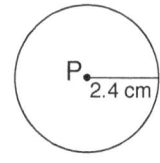

d. length of $\widehat{AB}$

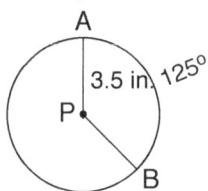

e. i) Area of inner circle P
   ii) Area of outer circle P

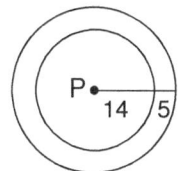

f. Area of circle P

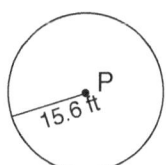

28. Find the shaded and unshaded areas of each given circle. Round to the nearest tenth if necessary.
«See Examples 10-12 to 10-16»

a.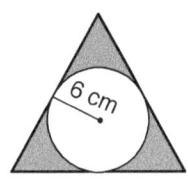

b.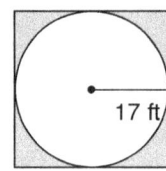

c.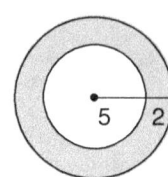

d.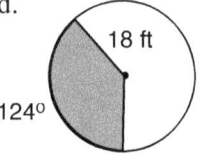

e.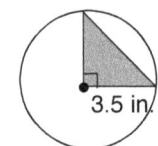

f.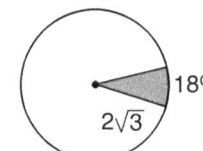

g.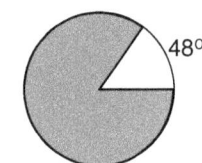

h.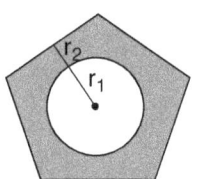
radius of a circle (r_1) = 16 ft
radius of a pentagon (r_2) = $16\sqrt{3}$ ft

i.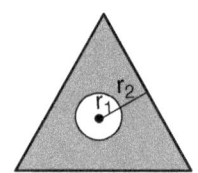
r_1 = 6 in.
r_2 = 18 in.

29. Solve the following problems.
«See Examples 10-12 to 10-16»

a. Find the area of a circle with a circumference of 17.4π cm.

b. Find the area of a circle with a diameter of 20 yards.

c. Find the area of a sector that measures 35° if the circle has an area of 48π cm^2.

d. Find the area of a circle with a sector of 78° if the sector has an area of 112π ft^2.

30. Find the area of the shaded regions. Round your answers to the nearest tenth.
«See Examples 10-12 to 10-16»

a.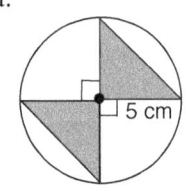

b.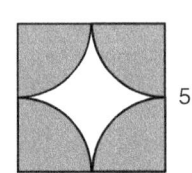

c.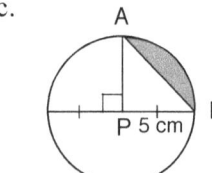

ANSWERS

(1) a. A polygon with all its interior angles has less than 180°, b. At least one angle in a polygon is more than 180°, c. The angles and sides are congruent., d. n-gon = $(n - 2)(180°)$, e. n-gon(each exterior angle) = $((n - 2)(180°))/n$, 180°

(2) a. 900°, b. 2,340°, c. 4,140°, d. 360°, e. 1,800°, f. 6,120°, g. 180°, h. 540°

(3) a. 128°, b. 140°, c. 139°, d. 185°, e. 153°

(4) a. 133°, b. 135°, c. 128°

(5) a. 60°, b. 27.69°, c. 30°, d. 15°, e. 45°, f. 36°

(6) a. 120°, b. 51.4°, c. 45°

(7) a. 2,160°, b. 1,800°, c. 2,880°, d. 32.7°, e. 1,080°

(8) a. $n = (306°/(180° - i))$, where n is the number of side and i is the each interior angle, 15 sides, b. 48 sides, c. 19 sides, d. 14 sides

(9) a. 114°, b. 42°, c. 32°, d. 43°, e. 25°, f. 17.6°

(10) a. 35.07 cm, b. 84.87 cm, c. 10.83 cm

(11) a. Octagon, b. 6 cm, c. AC, d. 12 cm, e. AD = 15.7, f. 96 cm, g. 45°, h. A = 696, i. 22.5°

(12) a. P = 70 units, A = 337.2 units^2, b. P = 48 units, A = 166.4 units^2, c. P = 72 units, A = 374.1 units^2, d. P = 88 units, A = 484 units^2, e. P = 40 units, A = 100 units^2, f. P = 31.2 units, A = 46.8 units^2

(13) a. P = 64 units, A = 256 units^2, b. P = 30 units, A = 65 units^2, c. P = 40 units, A = 110.1, d. P = 52 units, A = 130 units^2, e. P = 56 units, A = 196 units^2, f. P = 108 units, A = 841.8 units^2

(14) a. 824.2 units^2, b. 1349 units^2, c. 210.4 units^2, d. 753.5 units^2

(15) a. shaded = 421 units^2, unshaded = 596 units^2, b. shaded = 42.14 units^2, unshaded = 153.86 units^2, c. shaded = 93.5 units^2, unshaded = 19.5 units^2, d. shaded = 24.3 units^2, unshaded = 235.5 units^2, e. shaded = 19.55 units^2, unshaded = 6.9 units^2, f. shaded = 301 units^2, unshaded = 314 units^2

(16) a. 2/9 and 4/81, b. 1/6 and 1/36, c. 3.5/4.2 and 12.25/17.64

(17) a. 2:13, b. $\sqrt{6}$:2, c. $\sqrt{14}$:14, d. 1:7, e. 1:25, f. 1: 8, g. 1: 5.5, 18.84 : 44.18 or 1 : 2.35, h. 1:9

(18) a. $\sqrt{9.61\pi} / \sqrt{29.16\pi}$, b. $\sqrt{14\pi} / \sqrt{40.6\pi}$, c. $\sqrt{14.53} / \sqrt{90.8}$

(19) a. $\sqrt{10} / \sqrt{40}$, b. 96 ft^2, c. 249.6 ft^2, d. $\sqrt{24} / \sqrt{3}$

(20) a. 82°, b. 98°, c. 81.7 in., d. 18.6 in. e. 22.2 in. f. 40.8 in. g. 530.7 in.^{2}, h. 82°

(21) a. 15.7 in., b. 3.9 in, c. 2.5 in., d. 7.3 in., e. 33.6 in., f. 7.9 in.

(22) a. 18.84 cm, b. 15.7 in., c. 56.55 cm, d. 106.81 in, e. 50.27 cm, f. 3.8 cm

(23) a. 9.8 in, b. 20.4 cm, c. 0.8 cm, d. 2.3 cm, e. 1.9 cm, f. 6.1 in.

(24) a. 5.2 cm, b. 20 cm, c. 50.24 in., d. 43.96 cm, e. 6.8 units, f. 2.0 cm

(25) a. 69.1cm, b. 40.2 units, c. 6 units, d. 5.86 cm, e. 8 in., f. i) 62.8 units, ii) 81.64 units

(26) a. 88.9 ft^2, b. 410 ft^2, c. 121.2 in.^{2}, d. 334.6 ft^2, e. 9.8 in.^{2}, f. 49.6 ft^2, g. 115.4 in.^{2}, h. 76.9 in.^{2}, i. 153.9 in.^{2}

(27) a. 30.8 in., b. 8.2 units, c. 15.1 cm, d. 7.6 in., e. i) 615.4 units^2, ii) 1,133.5 units^2, f. 764.2 ft^2

(28) a. shaded = 74 cm^2, unshaded area = 113.1, b. shaded = 248.1 ft^2, unshaded area = 907.9 ft^2, c. shaded area = 75.4, unshaded area = 78.5, d. shaded area = 39 ft^2, unshaded area = 74.1 ft^2, e. shaded area = 6.1, unshaded area = 32.4, f. shaded area = 1.1, unshaded area = 36.7, g. shaded area = 38.1, unshaded area = 5.8, h. shaded area = 1983.3, unshaded area = 804.2, i. shaded area = 1570.5, unshaded area = 113.1

(29) a. 237.7 cm^2, b. 314 yd^2, c. 14.6 cm^2, d. 516.9π ft^2

(30) a. 25 cm^2, b. 19.6 in.^{2}, c. 7.1 cm^2

SELF-TEST

1. Which of the following statements defines the interior angles of a polygon?

    (a) The measure of each interior angle of a regular n-gon is $\frac{(n-2)(180°)}{n}$.
    (b) All interior angles of a regular n-gon are congruent.
    (c) The sum of the measures of the interior angles of a convex n-gon is $(n-2)(180°)$.
    (d) All of the above.

2. Which of the following statements does not define the exterior angles of a polygon?

    (a) The sum of the measures of the exterior angles of a convex polygon is 180°.
    (b) The measure of each exterior angle of a regular n-gon is $\frac{360°}{n}$.
    (c) The sum of the measures of the exterior angles of a convex polygon is 360°.
    (d) The measure of each exterior angle of a regular n-gon is $\frac{(n-2)(180°)}{n}$.

3. Which of the following is the sum of the interior angles of a 14-gon?

    (a) 360°
    (b) 180°
    (c) 2520°
    (d) 2160°

4. What is the sum of an interior angle of a regular 18-gon?

    (a) 360°
    (b) 180°
    (c) 2880°
    (d) 3240°

5. What is the sum of the exterior angle measures of a convex octagon?

    (a) 360°
    (b) 180°
    (c) 1080°
    (d) 1440°

6. If the measure of an interior angle of a regular polygon is 157.5°, how many sides does the polygon have?

    (a) 14
    (b) 18
    (c) 24
    (d) 28

7. What is the measure of the unknown interior angle of a hexagon given the rest of the interior angles?
    130°, 85°, 96°, 138°, 143°, and $x°$

    (a) 125°
    (b) 128°
    (c) 150°
    (d) 145°

8. The measure of each interior angle of a regular polygon is 172.5°. How many sides does the polygon have?

    (a) 40
    (b) 45
    (c) 48
    (d) 50

9. The sum of the measure of the interior angles of a convex polygon is 3060°. How many sides does the polygon have?

    (a) 15
    (b) 18
    (c) 19
    (d) 20

10. Find the value of x given the diagram.

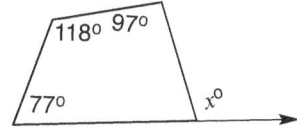

(a) 112°
(b) 110°
(c) 118°
(d) 120°

11. Which of the following is the sum of the exterior angles for a 20-gon?

(a) 180°
(b) 360°
(c) 162°
(d) 3240°

12. The sum of the measures of the interior angles of a convex pentagon is 540°. Which of the following is the value of x?

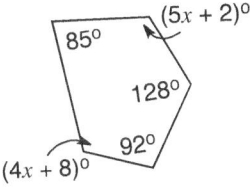

(a) 22
(b) 23
(c) 24
(d) 25

13. Which of the following statements does not define the area of an equilateral triangle?

(a) All side lengths are congruent.
(b) The area of an equilateral triangle is $\frac{1}{4}s^2\sqrt{3}$, where s is the side length.
(c) The area of an equilateral triangle with 2 cm sides is $\sqrt{3}$ cm².
(d) The area of an equilateral triangle is $\frac{1}{2}bh$, where b is the base length and h is the height.

14. Which of the following statements does not define the area of a regular polygon?

(a) The area of a regular n-gon with side length s is half the product of the apothem a and the perimeter P, $A = \frac{1}{2}aP$.
(b) The central angle of the triangle is 120°.
(c) The apothem (a) of the square is $\frac{1}{2}s$, where s is the side length.
(d) The apothem (a) of the hexagon is $\sqrt{3}s$, where s is the side length.

15. Find the area of each equilateral triangle given. Round to the nearest hundredth.

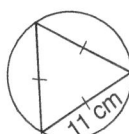

(a) 30.25 cm²
(b) 209.58 cm²
(c) 52.39 cm²
(d) 332.75 cm²

16. In the diagram, the side of a regular pentagon is 8 units. Find the length of GF.

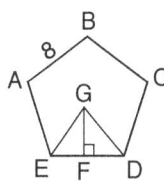

(a) 4.9
(b) 11.0
(c) 5.5
(d) 6.8

17. In the diagram, the length of the apothem of the hexagon is $2\sqrt{3}$ units. What is the area of the hexagon?

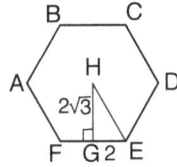

(a) $24\sqrt{3}$ units^2
(b) $12\sqrt{3}$ units^2
(c) $6\sqrt{3}$ units^2
(d) $3\sqrt{3}$ units^2

18. Which of the following statements does not define the apothem (a) of a regular polygon?

(a) The apothem (a) of a hexagon is $\frac{1}{2}\sqrt{3}s$, where s is the side length.
(b) The apothem (a) of a pentagon is $\frac{s}{2\tan(36°)}$, where s is the side length.
(c) The apothem (a) of a octagon is $\frac{s}{2\tan(22.5°)}$, where s is the side length.
(d) The apothem (a) of a hexagon is $\sqrt{3}s$, where s is the side length.

19. In the diagram, the length of HF is 8 units. Find the area of the hexagon.

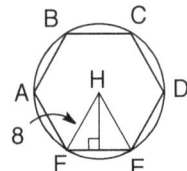

(a) 164.0 units
(b) 182.6 units
(c) 225.7 units
(d) 365.2 units

20. In the diagram, the radius of the circle is $6\sqrt{3}$. Find the area of ABCD.

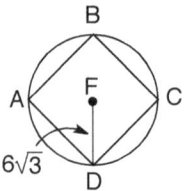

(a) 54.0 units
(b) 216.0 units
(c) 108.0 units
(d) 432.0 units

21. Which of the following statements does not define the area of similar polygons?

(a) If the ratio of the areas of two similar hexagons is a : b, then the ratio of their perimeter is a^2 : b^2.
(b) If the ratio of the perimeters of two similar hexagons is 2 : 5, then the ratio of their areas is 4 : 25.
(c) If two polygons are similar with the lengths of corresponding sides in the ratio of a : b, then the ratio of their areas is a^2 : b^2.
(d) If the ratio of the area of two similar kites is 3 : 2, then the ratio of their lengths of corresponding sides is $\sqrt{3} : \sqrt{2}$.

22. There are two similar pentagons with areas 24 cm^2 and 156 cm^2. Find the similarity ratio of the pentagons.

(a) 24 : 156
(b) 1 : 6.5
(c) 6 : 39
(d) $2\sqrt{6} : 2\sqrt{39}$

23. If the ratio of the areas of two similar triangles is 6 : 4, find the ratio of the lengths of the corresponding sides.

(a) 6 : 4
(b) 3 : 2
(c) 1.5 : 1
(d) $\sqrt{6} : 2$

24. If two similar triangles have areas of 8 in.² and 48 in.², find the similarity ratio of each pair of triangles.

(a) 64 : 2304
(b) $2\sqrt{2} : 4\sqrt{3}$
(c) 4 : 24
(d) 2 : 12

25. The areas of two similar trapezoids are 4 ft² and 32 ft². What is the similarity ratio of the lengths of the corresponding sides?

(a) 4 : 32
(b) 1 : 8
(c) $2 : 4\sqrt{2}$
(d) 2 : 16

26. The areas of two similar circles are 9π cm² and 49π cm². What are the similarity ratios of the areas and perimeters?

(a) 9 : 49 and $1 : \frac{49}{9}$
(b) 3 : 7 and $1 : \frac{7}{3}$
(c) $1 : \frac{49}{9}$ and $1 : \frac{49}{9}$
(d) 9 : 49 and 3 : 7

27. The ratio of the two similar circles is shown below. What is the ratio of the lengths of the circles A and B?

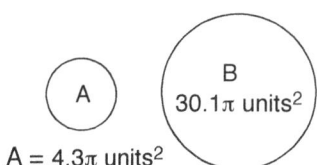
A = 4.3π units²   B 30.1π units²

(a) 4.3 : 30.1
(b) 1 : 7
(c) 1 : 7
(d) $\sqrt{4.3} : \sqrt{30.1}$

28. The ratio of the two similar hexagons is shown below. What is the ratio of the areas of the hexagons A and B?

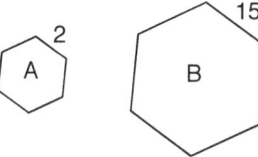

(a) $2^2 : 15^2$
(b) $\frac{15}{2}$
(c) $\frac{2}{15}$
(d) 2 : 15

29. The ratio of the two similar triangles is shown below. What is the ratio of the side lengths? What is the ratio of the areas of the triangles A and B?

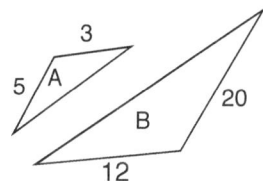

(a) 3 : 12 and $\sqrt{3} : 2\sqrt{3}$
(b) 3 : 12 and 1 : 4
(c) 1 : 4 and $1 : \sqrt{4}$
(d) 1 : 4 and 1 : 16

30. Which of the following statement does not define the circumference of a circle?

(a) The ratio of the circumference to the diameter is π.
(b) The circumference C of a circle is $C = 2\pi r$, where r is the radius of the circle.
(c) If the diameter of the circle is 3 cm, then the circumference of this circle is $2\pi(1.5)$.
(d) If the ratio of the diameter of two similar circles is a : b, then the ratio of their circumferences is a² : b².

31. Which of the following statements does not define an arc length?

(a) In a circle, the ratio of the length of a given arc to the circumference is equal to the ratio of the arc measure to 360°.
(b) $\frac{Arc\ length\ \widehat{AB}}{2\pi r} = \frac{m\widehat{AB}}{360°}$.
(c) A portion of the circumference of a circle.
(d) If the ratio of the arc lengths AB of two similar circles is a : b, then the ratio of their circumferences is $a^2 : b^2$.

32. Find the radius of the circle that has a circumference of 10.4π cm.

(a) 10.4 cm
(b) 5.2 cm
(c) 20.8 cm
(d) $\sqrt{10.4}$ cm

33. Find the circumference that has a radius of 8 in. Round your answer to the nearest tenth (π = 3.14).

(a) 25.1 in.
(b) 50.2 in.
(c) 12.6 in.
(d) $\sqrt{8}\,\pi$ in.

34. Find the arc length of a circle that has a radius of 6 ft and an arc angle of 65°. Round your answer to the nearest tenth (π = 3.14).

(a) 6.8 ft
(b) 13.6 ft
(c) 3.4 ft
(d) 1.4 ft

35. Find the measure of an arc angle of a circle given the diameter is 10 cm and the arc length is 2 cm. Round your answer to the nearest tenth (π = 3.14).

(a) 20°
(b) 21°
(c) 22°
(d) 23°

36. The arc measure is 69° and the length of the diameter is 12 in. What is the area of the sector of the circle below? Round your answer to the nearest tenth.

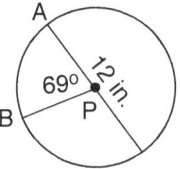

(a) 14.4 ft²
(b) 21.7 ft²
(c) 7.2 ft²
(d) 86.7 ft²

37. The measure of the given arc angle is 171° and the length of the radius is 3 cm. What is the area of the sector of the circle below? Round your answer to the nearest tenth.

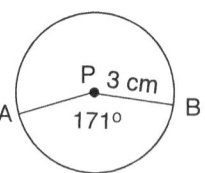

(a) 17.9 cm²
(b) 13.4 cm²
(c) 8.9 cm²
(d) 26.8 cm²

38. The arc measure is 43° and the circumference of the circle is 3.6π cm². What is the area of the sector below? Round your answer to the nearest tenth.

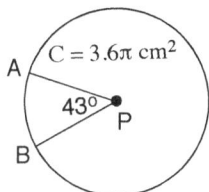

(a) 1.2 cm²
(b) 2.4 cm²
(c) 1.4 cm²
(d) 2.8 cm²

39. The measure of the given arc angle is 46° and the length of the radius is 11 cm. What is the arc length of the circle below? Round your answer to the nearest tenth.

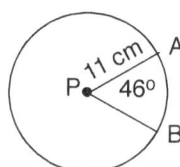

(a) 17.7 cm
(b) 8.8 cm
(c) 4.4 cm
(d) 48.5 cm

40. The measure of the given arc angle is 38°. What is the arc length of the circle below? Round your answer to the nearest tenth.

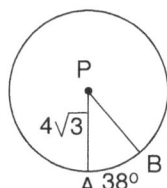

(a) 11.5 units
(b) 9.2 units
(c) 4.6 units
(d) 2.3 units

41. Which of the following statements does not define the area of a circle?

(a) The area of a circle is π times the square of the radius.
(b) The formula of the area of a circle is $A = \pi r^2$, where r is the radius of the circle.
(c) If the ratio of the radius of two similar circles is a : b, then the ratio of their areas is $a^2 : b^2$.
(d) If the radius of the circle is 2 cm, then the circumference of this circle is $2\pi(2)$.

42. Which of the following statements does not define the area of a sector of a circle?

(a) If the radius of the circle is 2 cm with the arc AB = 72°, then the area of the sector is 0.8π cm².
(b) The formula of the area of a sector of a circle is $A = \dfrac{m\widehat{AB}}{360°} \pi r^2$, where r is the radius.
(c) The ratio of the area of a sector to the area of the circle is equal to the ratio of the intercepted arc measure to 360°.
(d) If the ratio of the radius of two similar circles is a : b, then the ratio of the areas of their sectors is $a^2 : b^2$.

43. The measure of the given arc angle is 23° and the diameter of the circle is 17 ft. What is the indicated dot area of the sectors? Round your answer to the nearest tenth.

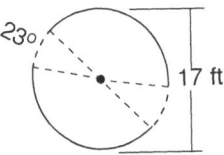

(a) 14.5 ft²
(b) 29.0 ft²
(c) 3.4 ft²
(d) 6.8 ft²

44. The measure of the given arc angle is 87° and the length of the radius is 15 ft. What is the indicated dot area of the sector of the circle below? Round your answer to the nearest tenth.

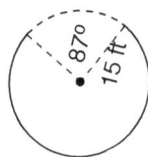

(a) 45.6 ft²
(b) 22.8 ft²
(c) 11.4 ft²
(d) 170.7 ft²

45. The arc measure is 252° and the length of the radius is 6 in. What is the indicated dot area of the sector? Round your answer to the nearest tenth.

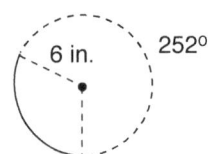

(a) 6.6 in.²
(b) 26.4 in.²
(c) 13.2 in.²
(d) 79.1 in.²

46. What is the arc length of the circle below? Round your answer to the nearest tenth.

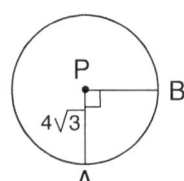

(a) 10.9 units
(b) 21.8 units
(c) 5.4 units
(d) 37.7 units

47. Use the diagram to find the area of the circle. Round to the nearest tenth.

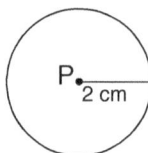

(a) 3.1 cm²
(b) 6.3 cm²
(c) 12.6 cm²
(d) 25.1 cm²

48. Use the diagram to find the area of the outer circle. Round to the nearest tenth.

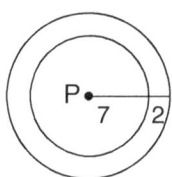

(a) 12.6 units²
(b) 100.5 units²
(c) 254.3 units²
(d) 153.9 units²

49. Find the shaded area of the triangle below. If necessary, round your answer to the nearest tenth.

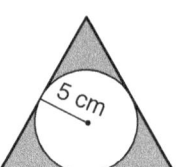

(a) 259.8 cm²
(b) 78.5 cm²
(c) 181.3 cm²
(d) 51.4 cm²

50. Find the shaded area of the square below. If necessary, round your answer to the nearest tenth.

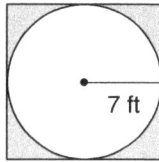

(a) 152.0 ft²
(b) 42.1 ft²
(c) 196.0 ft²
(d) 153.9 ft²

51. Find the area of the shaded sector. If necessary, round your answer to the nearest tenth.

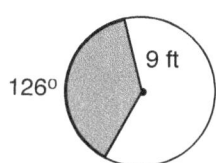

(a) 89.0 ft²
(b) 178.0 ft²
(c) 165.3 ft²
(d) 254.3 ft²

52. Find the area of the unshaded area. If necessary, round your answer to the nearest tenth.

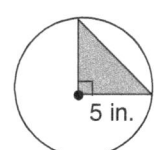

(a) 66.0 in.²
(b) 132.0 in.²
(c) 12.5 in.²
(d) 78.5 in.²

53. Find the area of the unshaded area. If necessary, round your answer to the nearest tenth.

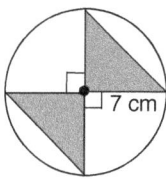

(a) 153.9 cm²
(b) 98.0 cm²
(c) 49.0 cm²
(d) 104.9 cm²

54. Find the shaded area of the circle below. If necessary, round your answer to the nearest tenth.

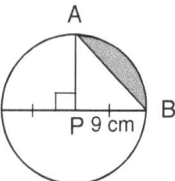

(a) 23.1 cm²
(b) 40.5 cm²
(c) 63.6 cm²
(d) 81.0 cm²

55. Find the unshaded area of the figure below. If necessary, round your answer to the nearest tenth.

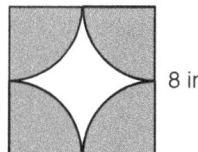

(a) 64.0 in.²
(b) 50.2 in.²
(c) 13.8 in.²
(d) 32.0 in.²

56. Find the area of a circle given the circumference is 17.4π cm.

(a) 54.6 cm²
(b) 475.3 cm²
(c) 950.7 cm²
(d) 237.7 cm²

57. Find the circumference of a circle given the area is 48π cm².

(a) 87.0 cm
(b) 43.5 cm
(c) 21.8 cm
(d) 10.9 cm

58. Find the area of a circle if a sector with a measure of 78° has an area of 112π ft².

(a) 129.2π ft²
(b) 258.5π ft²
(c) 516.9π ft²
(d) 1033.8π ft²

ANSWERS

(1) d	(2) a	(3) d	(4) c	(5) a	(6) c
(7) b	(8) c	(9) d	(10) a	(11) b	(12) d
(13) d	(14) d	(15) c	(16) c	(17) a	(18) d
(19) b	(20) b	(21) a	(22) d	(23) d	(24) b
(25) c	(26) d	(27) d	(28) a	(29) d	(30) d
(31) d	(32) b	(33) b	(34) a	(35) d	(36) b
(37) b	(38) a	(39) b	(40) c	(41) d	(42) d
(43) b	(44) d	(45) d	(46) a	(47) c	(48) b
(49) d	(50) b	(51) a	(52) a	(53) c	(54) a
(55) c	(56) d	(57) b	(58) c		

CHAPTER 11
Surface Area and Volume of Solids

In this chapter, you will find classify the different kinds of solids. You will encounter the various surface areas of solids, such as pyramids, prisms, and spheres and also be able to find their volume, and moreover, will find the surface area and volume of similar solids by using the scale factor.

CONCEPTS   **EXAMPLES**   **FORMULAS**   **VOCABULARY**

11-1. Find the number of vertices (V), faces (F), and edge (E) of the figure.

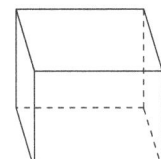

V = __8__

F = __6__

E = __12__

SOLUTION

By Euler's Theorem, the number of faces (F), vertices (V), and edges (E) of a polyhedron can be found by using the formula F + V = E + 2.

11-2. Formulas of surface areas and lateral areas of solids.

Name of Solid	Lateral Area (L.A.)		Surface Area (S.A.)	
Right Prism	L.A. = $p h$	p = perimeter h = height	S.A. = L.A. + 2B S.A. = $ph + 2(lw)$	B = area of a base B = lw l = length w = width
Cylinder	L.A. = Ch or L.A. = $2\pi rh$ C = $2\pi r$	C = circumference, h = height, r = radius	S.A. = L.A. + 2B S.A. = $2\pi rh + 2\pi r^2$	B = πr^2
Pyramid	L.A. = $\frac{1}{2}p\ell$ $\ell = \sqrt{a^2 + b^2}$	ℓ = slant height a = short leg b = long leg	S.A. = L.A. + B S.A. = $\frac{1}{2}p\ell + \frac{1}{2}ap$	Regular pyramid B = $\frac{1}{2}ap$ a = apothem p = perimeter
Cone	L.A. = $(\pi r\ell)$ $\ell = \sqrt{a^2 + b^2}$	ℓ = slant height	S.A. = L.A. + B S.A. = $\pi r\ell + \pi r^2$	B = πr^2
Sphere			S.A. = $4\pi r^2$	

11-3. What are the lateral and surface areas of a prism?

> **SOLUTION**
>
> The lateral area of a prism is the sum of the areas of the lateral faces. L.A. = *ph*, where *p* is the perimeter of the base and *h* is the height. The surface area of a prism is the sum of the areas of the faces. Therefore, S.A. = 2B + L.A., where B is the area of the base and L.A. is the lateral area.
>
>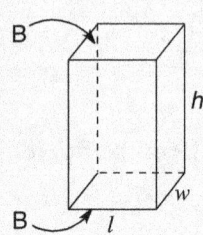
>
> L.A. = *ph*
> The lateral area of a prism if the sum of the areas of the faces without the area of the bases
>
> S.A. = 2B + *ph*
> 2B = B (top face of the prism) + B (bottom face of the prism)
> *p* = 2*l* + 2*w*   *p* = perimeter of the base
> *h* = height, *l* = length, and *w* = width

11-4. What are the lateral and surface areas of each triangular prism?

> The triangular prisms have three different shapes of the bases that are either right, equilateral, or regularly shaped triangular prisms. So finding the base area of the triangular prism can be calculated by using the different formulas of the bases. The surface area of the triangular prism is the sum of the area of the two bases and the lateral area.
>
> Right triangular prism   Equilateral triangular prism   Regular triangular prism
>
>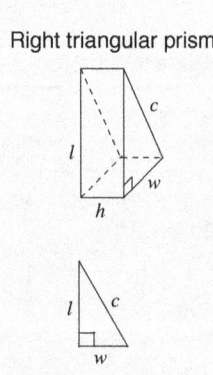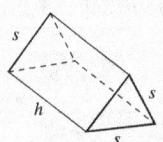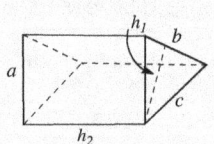
>
>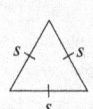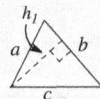
>
> B = (1/2)(*l*·*w*)   B = (1/4)(√3)(*s*²)   B = (1/2)(*b*)(*h₁*)
> L.A. = *ph* = (*l* + *w* + *c*)*h*   L.A. = *ph* = (*s* + *s* + *s*)*h*   L.A. = *ph* = (*a* + *b* + *c*)*h₂*
> S.A. = 2B + L.A.   S.A. = 2B + L.A.   S.A. = 2B + L.A.

11-5. What are the lateral and the surface areas of a cylinder?

SOLUTION

The lateral area of a cylinder is the product of the circumference of the base and the height. L.A. = Ch, where C is the circumference of the base and h is the height. $C = 2\pi r$, where r is the radius of a base. The surface area of a cylinder is the sum of lateral areas and the area of the two bases. Therefore, S.A. = 2B + L.A., where B is the area of the base and L.A. is the lateral area.

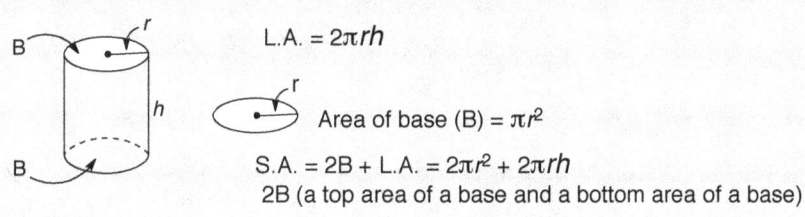

L.A. = $2\pi rh$

Area of base (B) = πr^2

S.A. = 2B + L.A. = $2\pi r^2 + 2\pi rh$
2B (a top area of a base and a bottom area of a base)

11-6. Find the lateral area and the surface area of a prism.

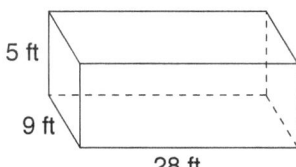

SOLUTION

By the definition of the surface area of a right prism, the surface area (S.A.) of a right prism can be found using the formula S.A. = 2B + L.A. The lateral area of a prism (L.A.) is the product of the perimeter and the height of the prism. L.A. = ph, where p is the perimeter of the base and h is the height.

L.A. = ph
L.A. = [2(28 ft) + 2(9 ft)](5ft) = (56 + 18)(5)
L.A. = 370 ft^2
The area of the base is B = lw = (9)(28) = 252 ft^2.
The perimeter of the base is $p = 2l + 2w$.
p = 2(28) + 2(9) = 56 + 18 = 74 ft.
So, S.A. = 2B + L.A. = 2(lw) + ph
  = (2)(252) + (74)(5)
  = 874 ft^2
So the lateral area of the prism is 375 ft^2 and the surface area of the prism is 874 ft^2.

11-7. Find the lateral area and the surface area of a prism.

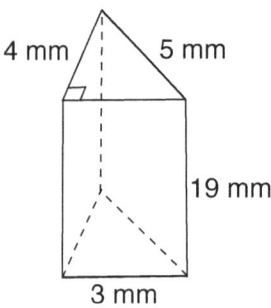

SOLUTION

By the definition of the surface area of a right prism, the surface area (S.A.) of a right prism can be found using the formula S.A. = 2B + L.A. The lateral area of a prism (L.A.) is the product of the perimeter of the base and the height of the prism. L.A. = ph, where p is the perimeter of the base, and h is the height. B = $\frac{1}{2}(lw)$, where B is the area of a base, l is the length of a base, and w is the width of a base.

L.A. = ph, p = 3 mm + 4 mm + 5 mm = 12 mm and h = 19 mm.
So the lateral area of a right prism is L.A = (12)(19) = 228 mm^2.
The surface area of a prism can be found by adding the area of the bases and the lateral area.

$$\text{S.A.} = 2B + \text{L.A.} = 2(\tfrac{1}{2})(lw) + ph$$

$$\text{S.A.} = 2(\tfrac{1}{2} \cdot 3 \cdot 4) + 228 = 240 \text{ mm}^2$$

So the lateral area of the right prism is 228 mm^2 and the surface area of the right prism is 240 mm^2.

11-8. Find the lateral area and the surface area of the cylinder. Put in terms of π.

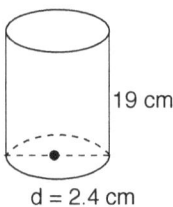

SOLUTION

By the definition of the surface area of a right cylinder, the surface area (S.A.) of a right cylinder is S.A. = 2B + L.A, where B is the area of a base. The lateral area of a cylinder (L.A.) is the product of the circumference of the base and the height. So L.A. = Ch, where C is the circumference of a base and h is the height. The circumference of the base of a cylinder is $2\pi r$.

L.A. = Ch = 2πrh
L.A. = 2π (1.2)(19)
= 45.6π cm²
S.A. = 2B + L.A = 2πr² + 2πrh
= 2π(1.2)² + 45.6π
= 48.48π cm²

So the lateral area of a cylinder is 45.6π cm² and the surface area of a cylinder is 48.48π cm².

11-9. Find the lateral area and the surface area of the pyramid.

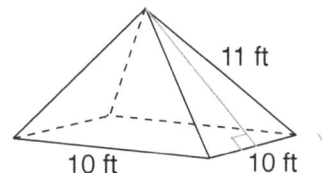

SOLUTION

By the definition of the surface area of a regular pyramid, the surface area (S.A.) of a regular pyramid is S.A. = B + L.A., where B is the area of the base. The lateral area of a pyramid (L.A.) is half the product of the perimeter and the slant height.

The perimeter of a pyramid in the given diagram can be found as (4 sides)(10 ft) = 4(10) = 40 ft. So L.A. = $\frac{1}{2}p\ell$, where p is the perimeter of a base and ℓ is the slant height.

L.A. = $\frac{1}{2}p\ell$	So, S.A. = B + L.A. = $s^2 + \frac{1}{2}p\ell$
= $\frac{1}{2}$(40)(11)	= $10^2 + \frac{1}{2}$(40)(11)
= 220 ft²	= 100 + 220
S.A. = B + L.A. The area of the base is B = s^2 = 10^2 ft² for the square.	= 320 ft²
	So the lateral area of a pyramid is 220 ft² and the surface area of the pyramid is 320 ft².

11-10. Find the lateral area and the surface area of the square pyramid.

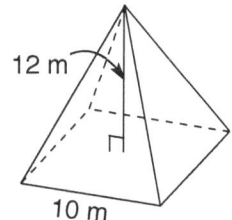

> **SOLUTION**
>
> By the definition of the surface area of a regular pyramid, the surface area (S.A.) of a regular pyramid is S.A. = B + L.A., where B is the area of the base. The lateral area of a pyramid (L.A.) is half the product of the perimeter and the slant height (ℓ).
>
> The sides of the base of the pyramid are congruent. You can find the slant height by using the Pythagorean Theorem.
>
>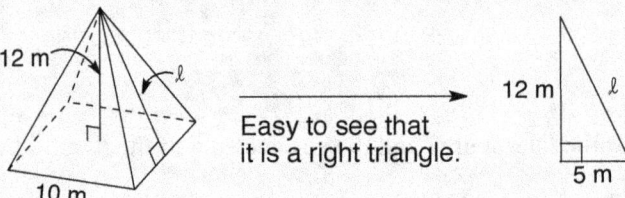
>
> $\ell = \sqrt{a^2 + b^2}$, where ℓ is the slant height, a is the short leg, and b is the long leg of the triangle.
> $\ell = \sqrt{(5)^2 + (12)^2}$
> $\ell = \sqrt{169} = 13$ m
> So the slant height of a pyramid is 13 m. Now you can find the lateral area in this pyramid.
> L.A. $= \frac{1}{2}p\ell$, where p is the perimeter of a base and ℓ is the slant height.
>
> $p = 4(10) = 40$ m.
> L.A. $= \frac{1}{2}p\ell = \frac{1}{2}(40)(13)$
> $\quad\quad = 260$ m^2
>
> S.A. = B + L.A. The area of the base is B = s^2 = 100 m^2 for the square.
> So, S.A. = $s^2 + \frac{1}{2}p\ell$,
> $\quad\quad = 100 + \frac{1}{2}(40)(13)$
> $\quad\quad = 360$ m^2
> So the lateral area of a pyramid is 260 m^2 and the surface area of the pyramid is 360 m^2.

11-11. Find the lateral area and the surface area of the regular polygon.

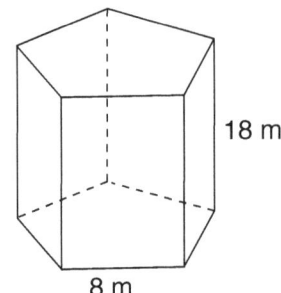

Chapter 11 Surface Area and Volume of Solids

SOLUTION

By the definition of the surface area of a regular prism, the surface area (S.A.) of a regular prism can be found using the formula S.A. = 2B + L.A. The lateral area of a prism (L.A.) is the product of the perimeter of the base and the height of the prism.

L.A. = ph, where p is the perimeter of the base, and h is the height.
L.A. = ph, $p = (ns) = 5(8) = 40$ m where n is the number of sides and s is the length of a side. The height is 18 m. So the lateral area of a right prism is L.A = $ph = (40)(18) = 720$ m². The formula for the base area of a regular polygon is $(1/2)aP$, where a is the apothem and P is the perimeter. $P = ns$, where n is the number of sides and s is the length of the side.

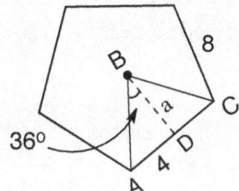

$$\tan(36°) = \frac{4}{a}$$
$$a = \frac{4}{\tan(36°)}$$
$$a = 5.5$$

$B = \frac{1}{2}a(ns) = \frac{1}{2}(5.5)(5 \cdot 8)$
$ = 110$ m².

The surface area of a prism can be found by adding the area of the bases and the lateral area.

S.A. = 2B + L.A. = $\frac{1}{2}a(ns) + ph$
S.A. = 2(110) + 720
S.A. = 940 m²

So the lateral area of the right prism is 110 m² and the surface area of the right prism is 940 m².

---

* For further information for the base in each polygon, review Chapter 10.
First find the top angle of the triangle in each polygon. The central angle of any polygon is then found by dividing the number of sides of an n-gon by 360°. So the central angle of a hexagon is 360°/6 = 60°. Afterwards, divide the central angle of the triangle to find the measure of ∠ABD or ∠CBD.

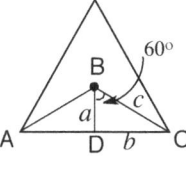

For a 30°-60°-90° triangle

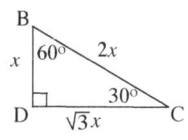

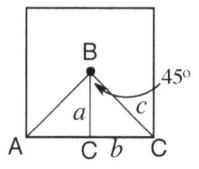

For a 45°-45°-90° triangle

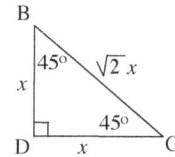

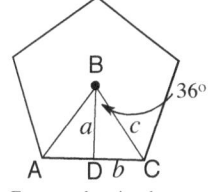

For a regular triangle

$a = (\cos 36°)(c)$
$b = (\sin 36°)(c)$
$b = (\tan 36°)(a)$

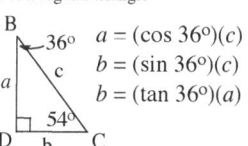

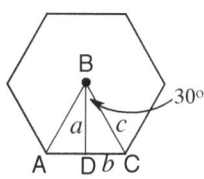

For a 30°-60°-90° triangle

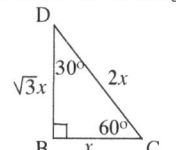

11-12. Find the lateral area and the surface area of the cone.

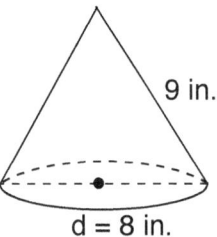

SOLUTION

By the definition of the surface area of a right cone, the surface area (S.A.) of a right cone is S.A. = L.A. + B, where B is the area of the base of the cone. The area of a base of a cone is πr^2 and the lateral area of a cone is $\pi r \ell$, where r is the radius of the base and ℓ is the slant height.

$$L.A. = \pi r \ell$$
$$= \pi(4)(9) = 36\pi \text{ in.}^2$$
$$S.A. = B + L.A. = \pi r^2 + \pi r \ell$$
$$= \pi(4^2) + \pi(4)(9)$$
$$= 52\pi \text{ in.}^2$$

So the lateral area is 36π in.^{2} and the surface area is 52π in.^{2}.

11-13. Find the surface area in the figure below.

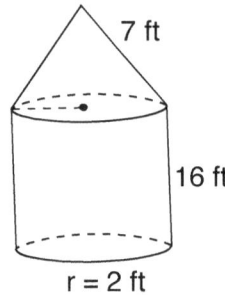

SOLUTION

You can find the sum of the regions by adding the lateral area of the right cone and the lateral area of cylinder. As it is collectively one solid, do not include the bottom of the cone or the top of the cylinder.

To find the surface area of the cone:
$$S.A. = L.A. + \text{no base}$$
$$= \pi r l = \pi(2)(7) = 14\pi \text{ ft}^2$$
$$= 43.96 \text{ ft}^2$$

To find the lateral area and a base area of the cylinder:
$$S.A. = L.A. + B = 2\pi r h + \pi r^2$$
$$= 2\pi(2)(16) + \pi(2^2)$$
$$= 68\pi \text{ ft}^2 = 213.52 \text{ ft}^2$$

Now you can add the two combined regions of the cone and cylinder.

Chapter 11 Surface Area and Volume of Solids

> S.A. = L.A.(cone) + L.A. (cylinder) + B (cylinder) = $\pi rl + 2\pi rh + \pi r^2$
> $= 14\pi \text{ ft}^2 + 64\pi \text{ ft}^2 + 4\pi \text{ ft}^2$
> $= 257.48 \text{ ft}^2$

11-14. Formula for finding the volume of solids.

Name of Solid	Volume (V)	
Right Prism	$V = Bh$ $B = lw$	B = area of the base, h = height, l = length, and w = width
Cylinder	$V = Bh$ $V = \pi r^2 h$	$B = \pi r^2$ B = area of the base h = height
Pyramid	$V = \frac{1}{3}Bh$	$B = lw$ B = area of the base h = height
Cone	$V = \frac{1}{3}Bh$, or $V = \frac{1}{3}\pi r^2 h$	$B = \pi r^2$
Sphere	$V = \frac{4}{3}\pi r^3$	

* When you find the volume of each solid, you should remember to use the height (h), NOT the slant height (ℓ).

11-15. Find the volume of the figure.

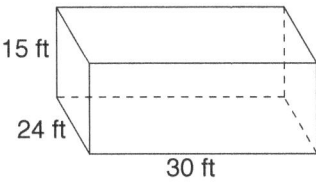

SOLUTION

By the definition of the volume of a prism, the volume V of a prism is $V = Bh$, where B is the area of a base and h is the height.
The area of the base of a prism is $B = lw$, where l is the length of the rectangle and w is the width.

So, $V = Bh = (lw)(h)$
$= (24)(30)(15) = 10{,}800 \text{ ft}^3$
Therefore, the volume of a prism is $10{,}800 \text{ ft}^3$.

11-16. Find the volume of the cylinder. Leave your answer in terms of π.

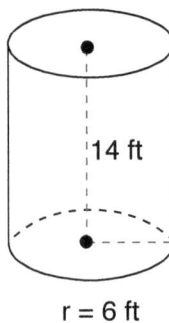

r = 6 ft

SOLUTION

By the definition of the volume of a cylinder, the volume V of a prism is V = Bh = $\pi r^2 h$, where B is the area of a base, h is the height, and r is the radius of a base.
The area of the base is B = πr^2 for a circle.
So, V = Bh = $\pi r^2 h$
= $(\pi)(6^2)(14)$
= 504π ft^3
Therefore, the volume of a cylinder is 504 ft^3.

11-17. Find the volume of an oblique cylinder. Leave your answer in terms of π.

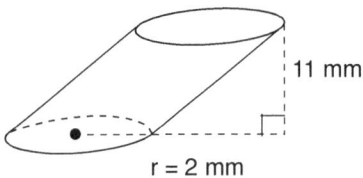

r = 2 mm

SOLUTION

Although the cylinder is oblique, it is still considered a right cylinder. So the formula for finding the area of a right cylinder can be still used.
So, V = Bh = $\pi r^2 h$
= $(\pi)(2^2)(11)$
= 44π mm^3

11-18. Find the volume of the outer cylinder to the nearest hundredth.

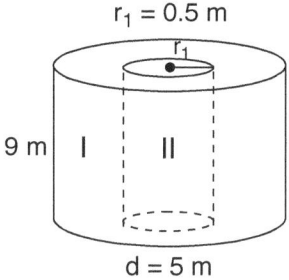

SOLUTION

Notice that inside the cylinder, there is an empty gap in the shape of another cylinder. So to find the volume of the outer cylinder, first find the volumes of the outer (I) and inner (II) cylinders.
To find the volume of the outer (I) cylinder.
$V = \pi r^2 h$, where r is the radius of the outer cylinder.
$= \pi(2.5)^2(9) = 56.25\pi$ ft^3
To find the volume of the inner (II) cylinder.
$V = \pi r^2 h$, where r_1 is the radius of the inner cylinder
$= \pi(0.5)^2(9) = 2.25\pi$ ft^3
You can now find the volume of the outer cylinder by subtracting the volume of the inner (II) cylinder from the volume of the outer (I) cylinder.
The volume of the outer cylinder = $56.25\pi - 2.25\pi = 54\pi$ ft^3

11-19. Find the volume of the square pyramid to the nearest whole number.

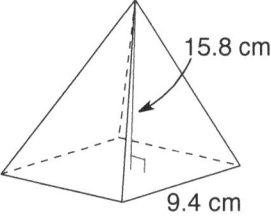

SOLUTION

By the definition of the volume of a pyramid, the volume V of a pyramid is $V = (1/3)Bh$, where B is the area of the base and h is the height.
The area of the base is $B = s^2$ for the square. $B = (9.4)(9.4) = 9.4^2$ cm^2.

$$\begin{aligned} \text{So, } V &= (1/3)Bh = (1/3)(lw)(h) \\ &= (1/3)(9.4^2)(15.8) \\ &\approx 465 \text{ cm}^3 \end{aligned}$$

11-20. Find the volume of the cone. Leave your answer in terms of π.

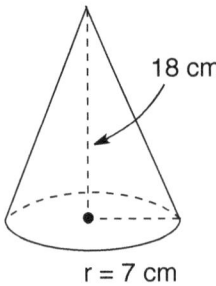

SOLUTION

By the definition of the volume of a cone, the volume V of a cone is $V = (\frac{1}{3})Bh = (\frac{1}{3})\pi r^2 h$, where B is the area of the base, h is the height, and r is the radius of the base.
The area of the base is $B = \pi r^2$ for a circle. $B = \pi(7^2)$ cm^2.
So, $V = (\frac{1}{3})Bh = (\frac{1}{3})(\pi r^2)h$
$= (\frac{1}{3})\pi(7^2)(18)$
$= 294\pi$ cm^3

11-21. Find the height of the cone. Round your answer to the nearest hundredth if necessary.

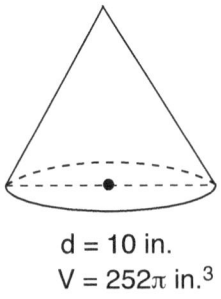

d = 10 in.
V = 252π in.^{3}

SOLUTION

By the definition of the volume of a cone, the volume V of a cone is $V = (\frac{1}{3})Bh = (\frac{1}{3})\pi r^2 h$, where B is the area of the base, h is the height, and r is the radius of the base.
So, $h = \frac{3V}{\pi r^2} = \frac{3(252\pi)}{\pi(5)^2} = \frac{756}{25} = 30.2$ in.
Therefore, the height of the cone is 30.2 in.

Chapter 11   Surface Area and Volume of Solids

11-22. Find the surface area of the sphere. Leave your answer in terms of π.

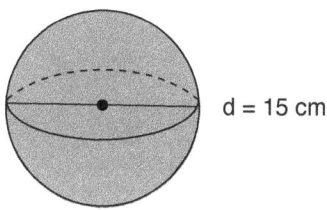

SOLUTION

By the definition of the surface area of a sphere, the surface area (S.A.) is S.A. $= 4\pi r^2$, where r is the radius of the sphere.
So, S.A. $= 4\pi r^2 = 4\pi(7.5)^2 = 225\pi$ cm^2
Therefore, the surface area of the sphere is 225π cm^2.

11-23. Find the surface area of the hemisphere. Leave your answer in terms of π.

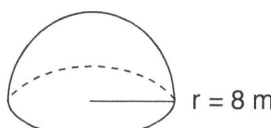

SOLUTION

The formula for the surface area of a sphere is S.A. $= 4\pi r^2$. Therefore, because a hemisphere is half of a sphere, the surface area of a hemisphere would be half of the surface area of a sphere.
$= (\frac{1}{2})4\pi(r)^2 = 2\pi r^2$
$= 2\pi(8)^2 = 128\pi$ m^2
Therefore, the surface area of the hemisphere is 128π m^2

11-24. Find the volume of the sphere. Leave your answer in terms of π.

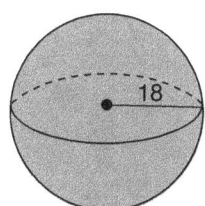

SOLUTION

By the definition of the volume of a sphere, the volume V of a sphere with radius r is $V = (4/3)\pi r^3$.
So, $V = (4/3)\pi r^3$
$= (4/3)\pi(18)^3$
$= 7776\pi$ units^3
Therefore, the volume of the sphere is 7776π units^3.

11-25. Find the volume of the unshaded region. Leave your answer in terms of π.

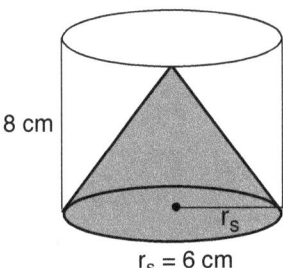

$r_s = 6$ cm

SOLUTION

Notice that the shaded region is in the shape of a cone, while the overall solid is a cylinder. The volume of a cone is $V = (1/3)Bh = (1/3)\pi r^2 h$, where B is the area of the base, h is the height, and r is the radius of the base. The volume of a cylinder is $V = Bh = \pi r^2 h$, where h is the height of the cylinder and r is the radius of the base.

To find the volume of a cone:
$V = (1/3)Bh = (1/3)\pi r^2 h$, where r is the radius of the cone.
$= (1/3)\pi(6)^2(8)$
$= 96\pi$ cm^3

To find the volume of a cylinder:
$V = Bh = \pi r^2 h$, where r is the radius of cylinder and h is the height of cylinder.
$= \pi(6)^2(8)$
$= 288\pi$ cm^3

You can now find the volume of the shaded region by subtracting the volume of the cone from the volume of the cylinder.
The volume of the unshaded region in the cylinder $= 288\pi - 96\pi = 192\pi$ cm^3.

PRACTICES

1. Name each figure.
《See Example 11-2》

a.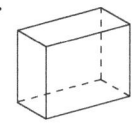

b.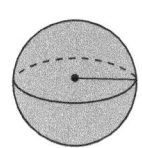

c.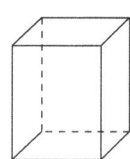

d.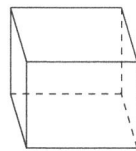

e.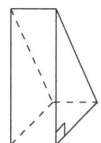

f.

g.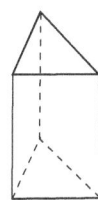

h.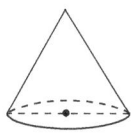

i.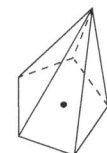

j.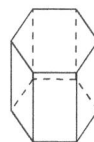

k.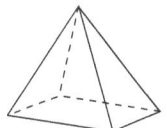

l.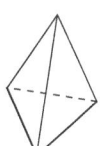

2. Find the number of vertices (V), faces (F), and edge (E) of each figure.
《See Example 11-1》

a. 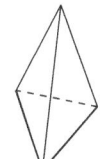 V = ____ F = ____ E = ____

b. 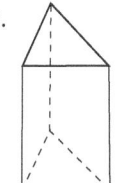 V = ____ F = ____ E = ____

c. 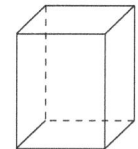 V = ____ F = ____ E = ____

150 Chapter 11 Surface Area and Volume of Solids

d. 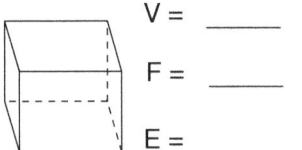 V = ____ F = ____ E = ____

e. 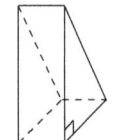 V = ____ F = ____ E = ____

f. 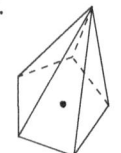 V = ____ F = ____ E = ____

g. 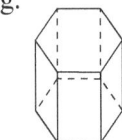 V = ____ F = ____ E = ____

h.  V = ____ F = ____ E = ____

i. V = ____ F = ____ E = ____

3. Solve each problem using the given information.
≪See Example 11-1≫

   a. A polyhedron has 6 vertices and 12 edges. Find the number of faces.

   b. A polyhedron has 20 faces and 30 edges. Find the number of vertices.

   c. A polyhedron has 12 faces and 20 vertices. Find the number of edges.

   d. A polyhedron has 26 faces and 18 vertices. Find the number of edges.

   e. A polyhedron has 6 faces and 12 edges. Find the number of vertices.

4. Find the lateral area of each prism.
≪See Examples 11-3 to 11-7≫

a.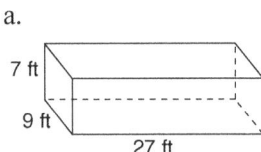
7 ft
9 ft
27 ft

b.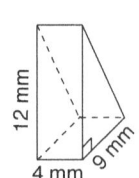
12 mm
4 mm
9 mm

5. Find the lateral and surface areas of each prism below.
«See Examples 11-3 and 11-6»

a.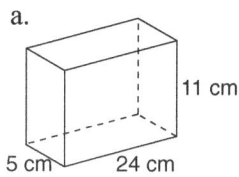

b.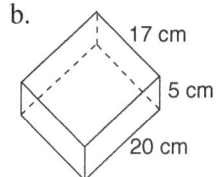

c.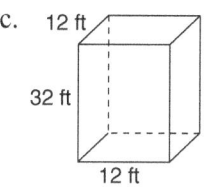

d.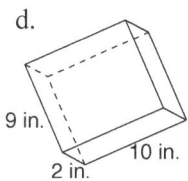

e.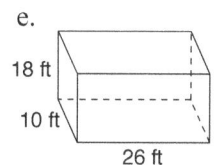

f.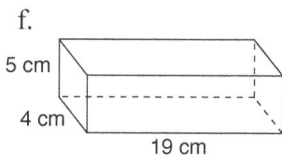

6. Solve each problem using the given information.
«See Examples 11-5 and 11-8»

   a. The lateral area of a cylinder is 5000 cm² with a radius of 36 cm. What is the height of the cylinder?

   b. The lateral area of a cylinder is 225 ft² with a diameter of 8 ft. What is the height of the cylinder?

   c. The radius of a cylinder has 17 ft. The height of this cylinder is 26 ft tall. What is the surface area of the cylinder?

   d. A cube with a base perimeter of 20 cm and a height of 7 cm. What are the lateral and surface areas of the cube?

   e. A rectangular prism with a height of 24 ft, a length of 6 ft, and a width of 7 ft. What is the surface area of the rectangular prism?

7. Find the lateral and surface areas of each prism.
≪See Examples 11-4 and 11-7≫

a.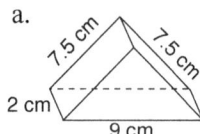

b.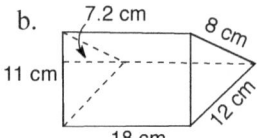

c.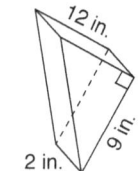

d.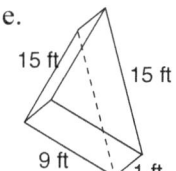

e.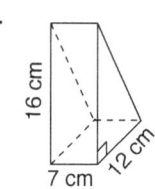

f.

8. Find the lateral and surface areas of each figure.
≪See Examples 11-2 to 11-7≫

a.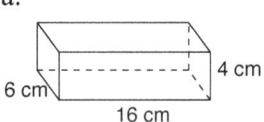

b.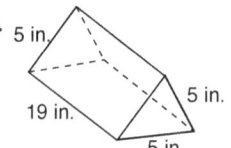

c.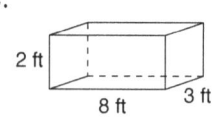

d.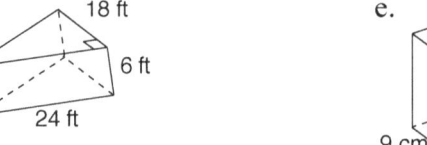

e.

f.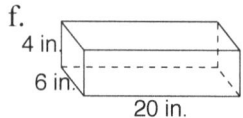

9. Find the lateral and surface areas of each cylinder. Give your answer in terms of π.
≪See Examples 11-5 and 11-8≫

a.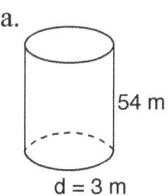

b.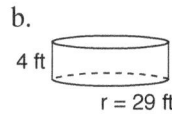

c.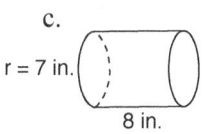

d.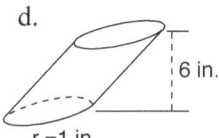

e.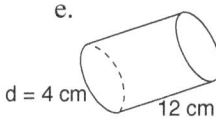

f.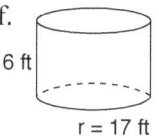

10. Find the lateral area of each square pyramid. Round your answer to the nearest whole number.
≪See Example 11-9≫

a.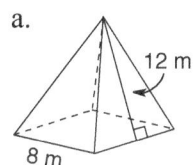

b.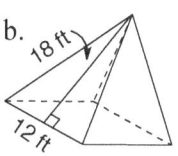

c.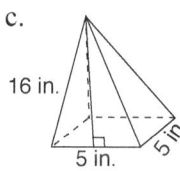

11. Find the lateral and surface area of each square pyramid. Round your answer to the nearest tenth if necessary.
≪See Examples 11-9 and 11-10≫

a.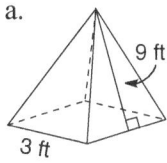

b.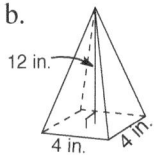

c.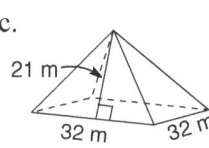

d.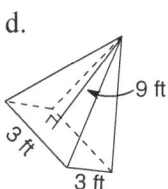

e.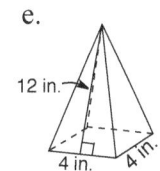

f.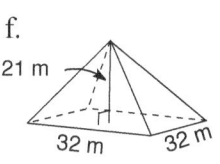

12. Solve the following problems.
≪See Examples 11-9 and 11-10≫

   a. Find the lateral area of a square pyramid that has a base side length of 9 cm and a slant height of 15 cm.

   b. Find the surface area of a square pyramid that has a lateral area of 30 ft^2 and a base area of 16 ft^2.

   c. A right cone has a radius of 8 in. and a slant height of 17 in. What is the surface area? Put your answer in terms of π.

   d. A cone has a diameter of 12 yd and a surface area of 423π yd^2. What is the length of the slant height?

13. Find the lateral area of each cone. Round your answer to the nearest whole number and give your answer in terms of π.
≪See Example 11-12≫

a.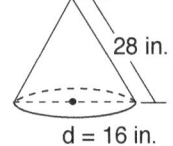

b.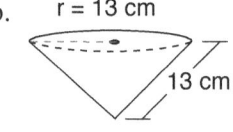

c.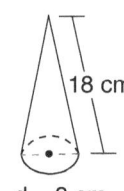

d.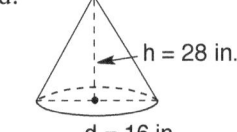

e.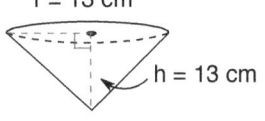

f.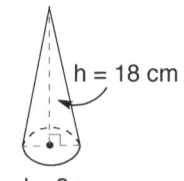

Chapter 11   Surface Area and Volume of Solids   155

14. Find the surface area of each right cone. Round your answer to the nearest whole number.
«See Example 11-12»

a. 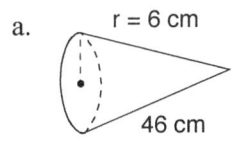   b. 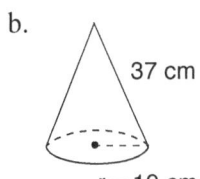   c.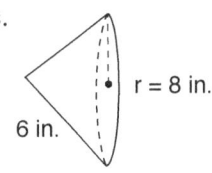

15. Find the surface area. Round your answer to the nearest whole number.
«See Examples 11-9 to 11-12»

a. 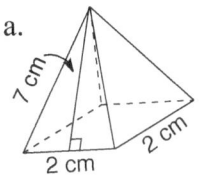   b. 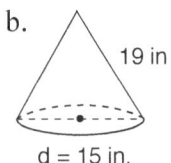   c.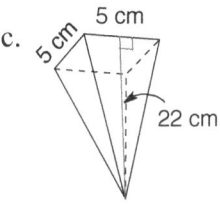

d. 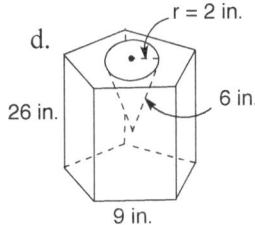   e. 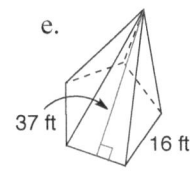   f.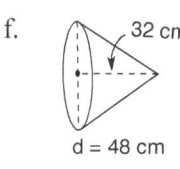

16. Solve each problem in the given information.
«See Examples 11-9 to 11-12»

a. The lateral area of a cone is 439.6 ft². The diameter of the base of the cone is 14 ft. What is the height of the cone?

b. A square pyramid has a slant height of 25 cm with a base perimeter of 20 cm. What is the surface area of the pyramid?

c. The height of a cone is 16 ft with a base diameter of 24 ft. What is the surface area of the cone?

d. The slant height of a square pyramid is 17 meters. The length of each side of the base is 8 meters. What is the surface area of the pyramid?

17. Find the surface area. Round your answer to the nearest whole number.
  《See Examples 11-9 to 11-13》

a.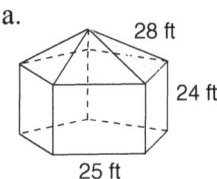

b.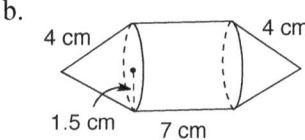

c.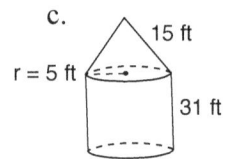

d.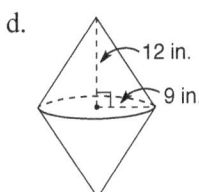

e.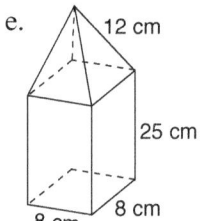

f.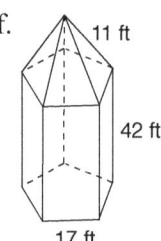

18. Find the volume of each prism. If necessary, round to the nearest tenth.
  《See Examples 11-14 and 11-15》

a.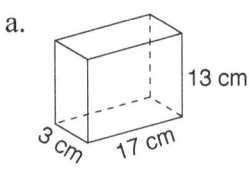

b.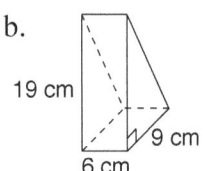

c.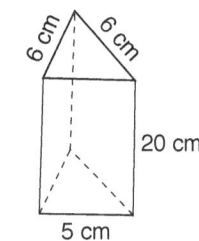

d.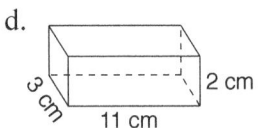

e.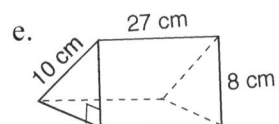

f.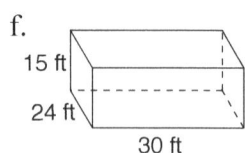

Chapter 11   Surface Area and Volume of Solids   157

19. Find the solution using the given information.
≪See Examples 11-14 and 11-15≫

   a. Find the volume of a rectangular prism that has a height of 32 cm, a length of 18 cm, and a width of 6 cm.

   b. Find the volume of a cube that has a side length of 7 in.

   c. Find the volume of a right rectangular prism that has a length of 15 mm, a width of 2 mm, and a height of 24 mm.

   d. Find the height of a rectangular prism that has a volume of 135 cm^3, a length of 9 cm, and a width of 3 cm.

   e. Find the length of a rectangular prism that has a volume of 250 ft^3, a height of 6 ft, and a width of 4 ft.

20. Find the volume of each following problem. If necessary, round to the nearest tenth.
≪See Example 11-15 and 11-16≫

   a.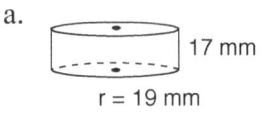
   b.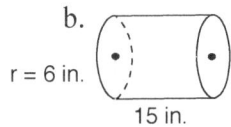
   c.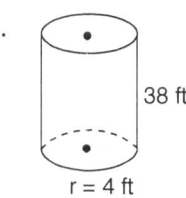

   d.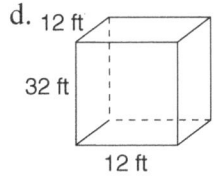
   e.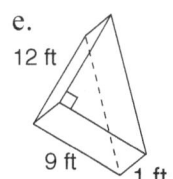
   f.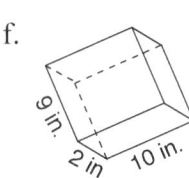

21. Find the volume of each figure. Round to the nearest tenth if necessary.
≪See Examples 11-15 to 11-17≫

   a.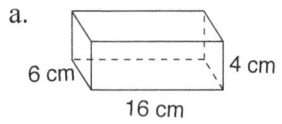
   b.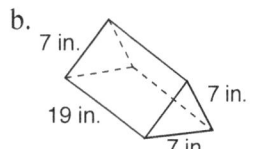
   c.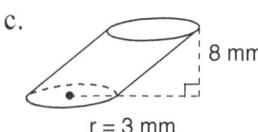

22. Find the solution using the given information.
    ≪See Examples 11-16 and 11-17≫

    a. The height of the cylinder is 19 ft. The radius of the base of the cylinder is 6 ft. What is the volume of the cylinder?

    b. The volume of a cylinder is 943 ft³. The diameter of the base of the cylinder is 15 ft. What is the height of the cylinder?

    c. The height of the cylinder is 24 cm. The diameter of the base of the cylinder is 9 cm. What is the volume of the cylinder?

    d. The volume of a cylinder is 141.3 m³. The height of the cylinder is 18 meters. What is the diameter of the base of the cylinder?

23. Find the volume of each cylinder. Round to the nearest tenth if necessary. Leave your answer in terms of π.
    ≪See Examples 11-16 and 11-17≫

    a.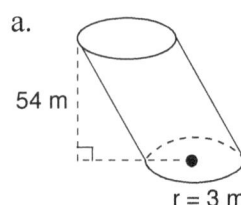

    b.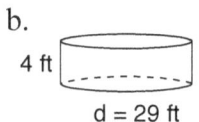

    c.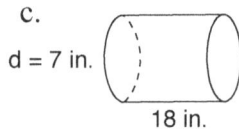

24. Find the volume of each figure. Round to the nearest tenth if necessary.
    ≪See Example 11-18≫

    a.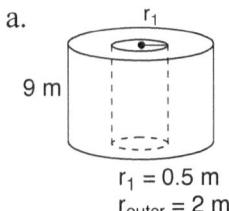

    b.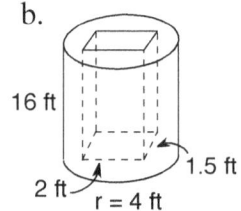

    c.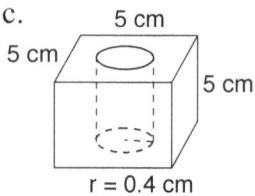

25. Find the volume of each pyramid. Round to the nearest whole number if necessary.
≪See Example 11-19≫

a.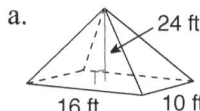

b.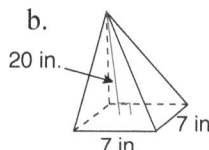

c.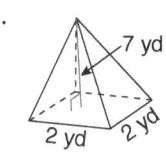

d.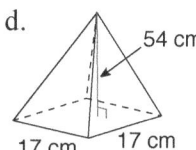

e.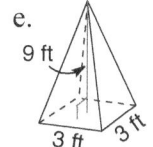

f.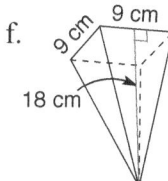

26. Find the solution using the given information.
≪See Example 11-19≫

   a. Find the volume of a square pyramid that has a height of 12 m and a length of 6 m.

   b. Find the volume of a right square pyramid that has a height of 8 ft and a base length of 3 ft.

   c. Find the volume of a regular pyramid that has a base area of 36 cm^2 and a height of 22.5 cm.

   d. Find the height of a right square pyramid. The volume of the pyramid is 1610 ft^3. The side length of the base is 6 ft.

27. Find the volume of each cone. Round to the nearest tenth if necessary.
≪See Examples 11-20 and 11-21≫

a.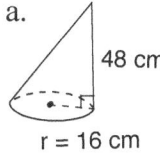

b.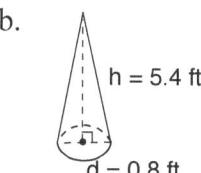

c.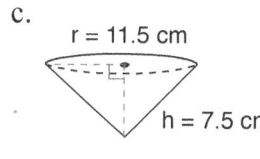

d.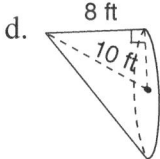

e.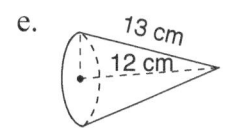

f.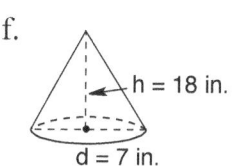

160   Chapter 11   Surface Area and Volume of Solids

28. Find the volume of each figure. If necessary, round to the nearest tenth.
≪See Examples 11-19 to 11-21≫

a.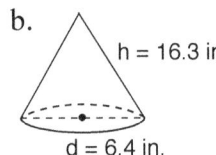

b.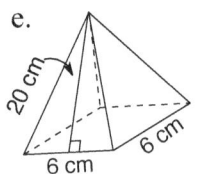

c. half cone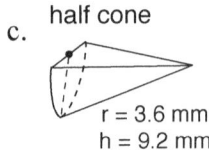

d.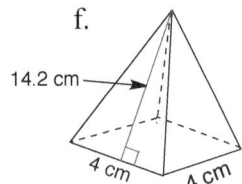

e.

f.

29. Find the volume of each given figure.
≪See Examples 11-18 to 11-20≫

    a. A cone has a height of 19 ft. The diameter of the cone is 6 ft. What is the volume of the cone?

    b. A right pyramid has a height of 28 m with a base area of 306.3 m². What is the volume of the pyramid?

    c. A cone has a radius of 10 ft with a height of 29 ft. What is the volume of the cone?

    d. A cone has a diameter of 3.5 mm and a height of 5.6 mm. What is the volume of the cone?

30. Find the volume. Round to the nearest whole number if necessary.
≪See Examples 11-19 to 11-21≫

a.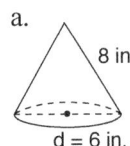

b.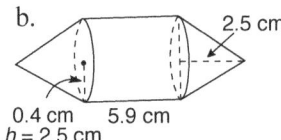

c.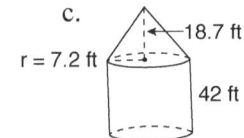

Chapter 11 Surface Area and Volume of Solids

d.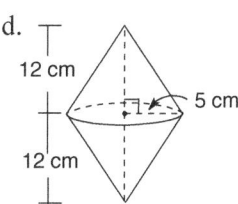

e.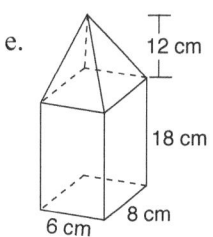

f.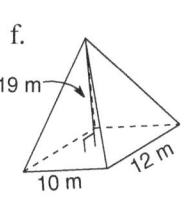

31. Find the surface area of the sphere. Give your answer in terms of π.
《See Examples 11-22 and 11-23》

a.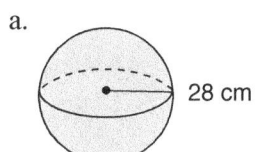

b.

c.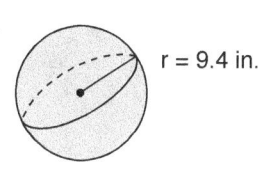

d.

e.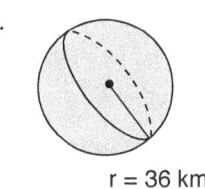

f.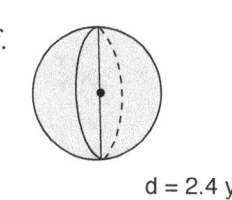

32. Find the solution using the given information.
《See Examples 11-22 and 11-23》

   a. Find the surface area of a sphere that has a radius of 12 in.

   b. Find the surface area of a basketball that has a diameter of 18 in.

   c. Find the diameter of a baseball that has a surface area of 62 in^2.

   d. Find the surface area of the moon, which has a diameter of 3500 km.

   e. A hemisphere has a diameter approximate to 6415 km. What is the surface area of the hemisphere?

33. Find the volume of the sphere. If necessary, round to the nearest tenth. Give your answer in terms of π.
   ≪See Example 11-24≫

   a.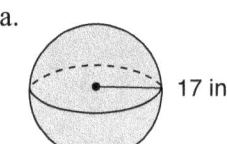

   b.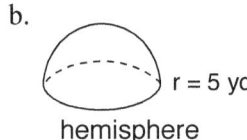

   c.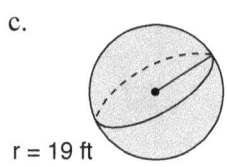

   d.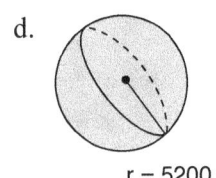

   e.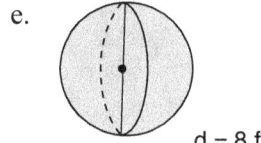

   f.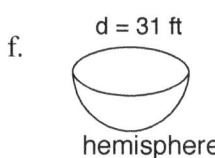

34. Find the surface area and the volume of the sphere. Give your answer in terms of π.
   ≪See Example 11-24≫

   a.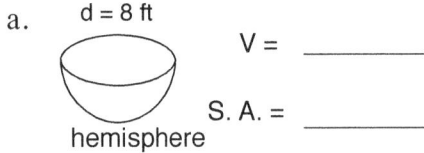

   b.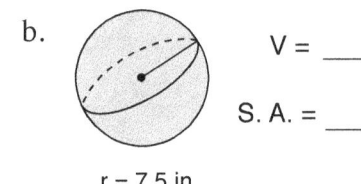

   c.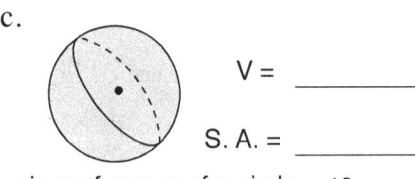

   d.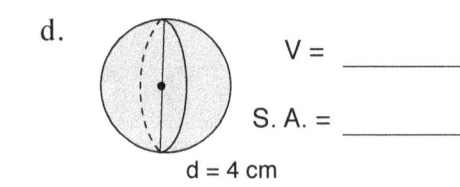

Chapter 11 Surface Area and Volume of Solids

35. Find each problem in the given information.
≪See Examples 11-22 and 11-24≫

    a. Find the volume and surface area of a sphere that has a radius of 7 yards.

    b. A sphere has a volume of 5461.3π. What is the diameter of the sphere?

    c. A softball has a radius of 4 inches. What is the volume of the softball?

    d. The surface area of a sphere is 4096π. What is the volume of the sphere?

    e. A hemisphere has a diameter of 3.8 in. What are the volume and surface area of the hemisphere?

36. Find the measure of the figure. If necessary, round to the nearest tenth. Give your answer in terms of π.
≪See Example 11-25≫

a.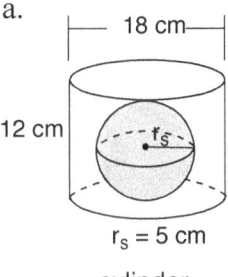

cylinder

  i. V_{sphere}

  ii. $V_{cylinder}$

  iii. $V_{cylinde} - V_{sphere}$

b.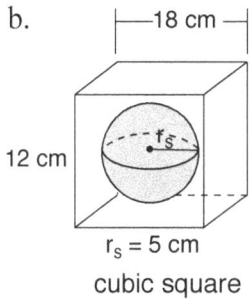

cubic square

  i. V_{sphere}

  ii. V_{box}

  iii. $V_{box} - V_{sphere}$

c.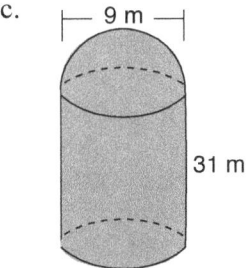

  i. $V_{hemisphere}$

  ii. $V_{cylinder}$

  iii. $V_{hemisphere} + V_{cylinder}$

d.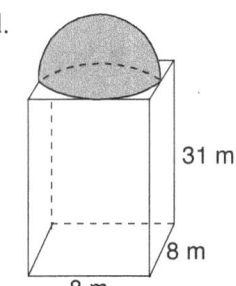

  i. $V_{hemisphere}$

  ii. V_{prism}

  iii. $V_{prism} + V_{hemisphere}$

ANSWERS

(1) a. rectangular prism, b. sphere,
c. rectangular prism, d. cube,
e. right triangular prism, f. cylinder,
g. triangular prism, h. cone,
i. pentagonal pyramid, j. hexagonal prism,
k. rectangular pyramid, l. triangular pyramid.

(2) a. V = 4, F = 4, E = 6, b. V = 6, F = 5, E = 9,
c. V = 8, F = 6, E = 12, d. V = 8, F = 6, E = 12,
e. V = 6, F = 5, = 9, f. V = 6, F = 6, E = 10,
g. V = 12, F = 8, E = 18, h. V = 5, F = 5, E = 8,
i. V = 5, F = 5, E = 9

(3) a. F = 8, b. V = 12, c. E = 30,
d. E = 42, e. V = 8

(4) a. 504 ft^2, b. 144 mm^2

(5) a. L.A. = 638 cm^2, S.A. = 878 cm^2,
b. L.A. = 370 cm^2, S.A. = 1050 cm^2,
c. L.A. = 1536 ft^2, S.A. = 1824 ft^2,
d. L.A. = 76 in.^{2}, S.A. = 256 in.^{2},
e. L.A. = 1296 ft^2, S.A. = 1816 ft^2,
f. L.A. = 230 cm^2, S.A. = 382 cm^2

(6) a. h = 22.1 cm,
b. h = 8.95 ft,
c. S.A. = 4593 ft,
d. L.A. = 140 cm^2, S.A. = 190 cm^2,
e. S.A. = 676 ft

(7) a. L.A. = 48 cm^2, S.A. = 102 cm^2,
b. L.A. = 558 cm^2, S.A. = 637.2 cm^2,
c. L.A. = 72 in.^{2}, S.A. = 180 cm^2,
d. L.A. = 600 cm^2, S.A. = 648 cm^2,
e. L.A. = 39 ft^2, S.A. = 147 ft^2,
f. L.A. = 336 cm^2, S.A. = 528 cm^2

(8) a. L.A. = 176 cm^2, S.A. = 368 cm^2,
b. L.A. = 285 in.^{2}, S.A. = 306.6 in.^{2},
c. L.A. = 44 ft^2, S.A. = 92 ft^2,
d. L.A. = 432 ft^2, S.A. = 864 ft^2,
e. L.A. = 1458 cm^2, S.A. = 4726 cm^2,
f. L.A. = 208 in.^{2}, S.A. = 448 in.^{2}

(9) a. L.A. = 162π m^2, S.A. = 166.5π m^2,
b. L.A. = 232π ft^2, S.A. = 1914π ft^2,
c. L.A. = 784π in.^{2}, S.A. = 896π in.^{2}
d. L.A. = 6π in.^{2}, S.A. = 6π in.^{2},
e. L.A. = 48π cm^2, S.A. = 56π cm^2,
f. L.A. = 204π ft^2, S.A. = 782π ft^2

(10) a. L.A. = 192 m^2, b. L.A. = 432 ft^2,
c. L.A. = 160 in.^{2}

(11) a. L.A. = 54 ft^2, S.A. = 63 ft^2,
b. L.A. = 96 in.^{2}, S.A. = 112 in.^{2},
c. L.A. = 1344 m^2, S.A. = 2368 m^2,
d. L.A. = 54.7 ft^2, S.A. = 63.7 ft^2,
e. L.A. = 97.3 in.^{2}, S.A. = 113.3 in^2,
f. L.A = 1689.6 m^2, S.A. = 2713.6 m^2

(12) a. L.A. = 270 cm^2, b. S.A. = 46 ft^2,
c. S.A. = 200π cm^2, d. l = 64.5 yd

(13) a. L.A. = 224π in.^{2}, b. L.A. = 169 π cm^2,
c. L.A. = 27π cm^2, d. 233π in.^{2},
e. 239π cm^2, f. 27π cm^2

(14) a. S.A. = 980 cm^2, b. S.A. = 1477 cm^2,
c. S.A. = 352 cm^2

(15) a. S.A. = 32 cm^2, b. S.A. = 624 in.^{2},
c. S.A. = 245 cm^2, d. S.A. for cone = 16π in.^{2}, S.A. for pentagon prism = 1448.7 in.^{2},
e. S.A. = 1920 ft^2, f. S.A. = 1536π cm^2

(16) a. h ≈ 18.7 ft, b. S.A. = 275 cm^2,
c. S.A. = 1205.76 ft^2, d. S.A. = 336 m^2

(17) a. S.A. = 5825.3 cm^2, b. S.A. = 33π cm^2,
c. S.A. = 410π ft^2, d. S.A. = 416π in.^{2},
e. S.A. = 1056 cm^2, f. S.A. = 4476.2 cm^2

(18) a. V = 663 cm^3, b. V = 513 cm^3,
c. V = 272.7 cm^3, d. V = 66 cm^3,
e. V = 648 cm^3, f. V = 10800 ft^3,

(19) a. V = 3,456 cm^3, b. V = 343 in.^{3},
c. V = 720 mm^3, d. h = 5 cm,
e. l = 10.4 ft

(20) a. V = 19,280 mm^3, b. V = 1696 in.^{3},
c. V = 1909 ft^3, d. V = 4,608 ft^3,
e. V = 54 ft^3, f. V = 180 in.^{3}

(21) a. V = 384 cm^3,
b. V = 403.1 cm^3,
c. V = 226.2 mm^2

(22) a. V = 2,148.8 ft^3, b. h = 5.3 ft,
c. V = 1526.8 cm^3, d. d ≈ 3.2 m

(23) a. V = 486π m^3, b. V = 841π ft^3,
c. V = 220.5π in.^{3}

(24) a. V($r_2$cylinder) = 113.1 m^3,
V($r_1$cylinder) = 0.9 m^3,
b. V(cylinder) = 803.9 ft^3,
V(prism) = 48 ft^3
c. V(cubic) = 125 cm^3,
V(cylinder) = 2.5 cm^3

(25) a. V = 1,280 ft^3, b. V = 327 in.^{3},
c. V = 9 yd^3, d. V = 5,202 cm^3,
e. V = 27 ft^3, f. V = 471 cm^3

(26) a. V = 120 m^3, b. V = 24 ft^3,
c. V = 270 cm^3, d. h = 134 ft
(27) a. V = 12868 cm^3, b. V = 0.9 ft^3,
c. V = 1038.7 cm^3, d. V = 301 ft^3,
e. V = 314.2 cm^3, f. V = 230.9 in.^{3}
(28) a. 1092.7 cm^3, b. 174.8 in^3,
c. 62.4 mm^3, d. 664 ft^3,
e. 237.3 cm^3, f. 75 cm^3
(29) a. V = 178.98 ft^3, b. V = 2,858.8 m^3,
c. V = 3,035.3 ft^3, d. V = 18.0 mm^3
(30) a. V = 70 in.^{3}, b. V = 4 cm^3,
c. V = 7852 ft^3, d. V = 1,256 cm^3,
e. V = 1056 cm^3, f. V = 760 m^3
(31) a. S.A. = 3136π cm^2,
b. S.A. = 9,523,396π mi^2,
c. S.A. = 353.4π in.^{2},
d. S.A. = 89.78π m^2,
e. S.A. = 5,184π km^2,
f. S.A. = 5.76π yd^2
(32) a. S.A. = 576π in.^{2},
b. S.A. = 324 π in.^{2},
c. d = 4.4 in.,
d. S.A. = 38,465,000 km^2,
e. S.A. = 29,642,404.63 km^2,
(33) a. V = 6,550.7π in.^{3},
b. V = 166.7π yd^3,
c. V = 9,145.3π ft^3,
d. V ≈ 1.9 x 10^{11}π mi^3,
e. V = 85.3π ft^3,
f. V = 2,482.6π ft^3
(34) a. V = 85,3π ft^3, S.A. = 32π ft^2,
b. V = 562.5π in.^{3}, S.A. = 225π in.^{2},
c. V = cm^3,
d. V = 10.7π cm^3, S.A. = 16π cm^2,
(35) a. V = 457.3π yd^3, S.A. = 196π yd^2,
b. V = cm^3, c. V = 85.3 in.^{3},
d. S.A. = 4,096π units^2,
e. V = 4.6π in.^{3}, S.A. = 7.22π in.^{2}
(36) a. i. 166.7π cm^3, ii. 972π cm^3,
iii. 2,528.6π cm^3,
b. i. 166.7π cm^3, ii. 3,888π cm^3,
iii. 3,364.6π cm^3,

SELF-TEST

1. How many edges does the polygon have?

   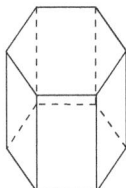

   (a) 22
   (b) 18
   (c) 12
   (d) 8

2. How many vertices does the polygon have?

   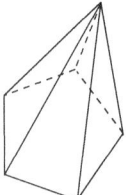

   (a) 6
   (b) 8
   (c) 10
   (d) 12

3. A polyhedron has 20 faces and 30 edges. Find the number of vertices that it has.

   (a) 52
   (b) 12
   (c) 32
   (d) 20

4. A polyhedron has 12 faces and 20 vertices. Find the number of edges that it has.

   (a) 24
   (b) 22
   (c) 32
   (d) 30

5. Which of the following statements are false?

   (a) The surface area of a prism is the sum of the areas of the bases and the lateral area
   (b) The surface area of a rectangular prism is the sum of twice the area of the base and the product of the base perimeter and the height, represented by 2B + Ph
   (c) The surface area of any right, equilateral, or regularly shaped triangular prism has the same formula for the area of the base.
   (d) The surface area of a right triangular prism is the sum of twice the area of the base and the product of the base perimeter and the height.

6. The side lengths of an equilateral triangular prism are 4 in. Which of the following is the area of the base of the equilateral prism? Round your answer to the nearest tenth if necessary.

   (a) $16\sqrt{3}$ in.^{2}
   (b) $4\sqrt{3}$ in.^{2}
   (c) 4.0 in.^{2}
   (d) 16.0 in.^{2}

7. Find the area of the base of the triangular prism. Round your answer to the nearest tenth if necessary.

   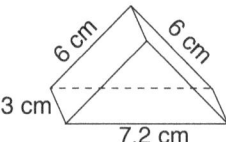

   (a) 17.3 cm^2
   (b) 48.1 cm^2
   (c) 46.1 cm^2
   (d) 60.5 cm^2

8. If a regular triangular prism has a base of 8 cm with a height of 6 cm, find the area of the base of the prism. Round your answer to the nearest tenth if necessary.

(a) 48.0 cm²
(b) 24.0 cm²
(c) 12.0 cm²
(d) 12√3 cm²

9. Find the lateral area of the right triangular prism. Round your answer to the nearest tenth if necessary.

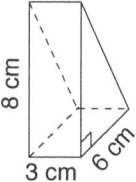

(a) 24.0 cm²
(b) 48.0 cm²
(c) 72.0 cm²
(d) 144.0 cm²

10. An isosceles triangular prism has two side lengths of 5 cm, a side length of 4, and a height of 6 cm. Find the lateral area of the triangular prism. Round your answer to the nearest tenth if necessary.

(a) 64 cm²
(b) 80 cm²
(c) 32 cm²
(d) 96 cm²

11. Given that a rectangular prism has a lateral area of 224 ft² and a base perimeter of 14 ft, find the height of the rectangular prism. Round your answer to the nearest tenth if necessary.

(a) 13 ft
(b) 14 ft
(c) 15 ft
(d) 16 ft

12. Find the lateral area of the prism. Round your answer to the nearest tenth if necessary.

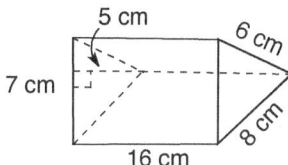

(a) 168.0 cm²
(b) 105.0 cm²
(c) 336.0 cm²
(d) 112.0 cm²

13. The lateral area of the rectangular prism is 120 ft² and the area of its base is 75 ft². Which of the following is the surface area of the rectangular prism?

(a) 195 ft²
(b) 225 ft²
(c) 250 ft²
(d) 315 ft²

14. Find the lateral area of the rectangular prism. Round your answer to the nearest tenth if necessary.

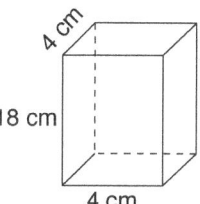

(a) 16.0 cm²
(b) 72.0 cm²
(c) 144.0 cm²
(d) 288.0 cm²

15. Which of the following statements are false?

(a) The lateral area of a cylinder is the product of the circumference and the height.
(b) The surface area of a cylinder is the sum of twice the area of the base and the product of the base perimeter and the height.
(c) The surface area of a cylinder is S.A. = $2\pi r^2 + 2\pi rh$
(d) The base area of a cylinder is $2\pi r^2$.

16. Which of the following is the area of the base of the cylinder?

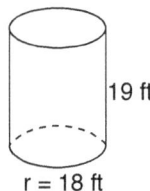

(a) 2147.8 ft²
(b) 1073.9 ft²
(c) 1017.4 ft²
(d) 2034.7 ft²

17. A cylinder has a height of 34 m and a diameter of 5 cm. Find the area of the base of the cylinder. Round to the nearest tenth.

(a) 573.1 cm²
(b) 533.8 cm²
(c) 39.3 cm²
(d) 157.0 cm²

18. Given that the height of a cylinder is 6 ft and the radius of its base is 31 ft, find the lateral area of the cylinder. Round to the nearest tenth.

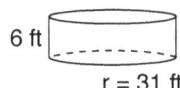

(a) 584.0 ft²
(b) 1168.1 ft²
(c) 3601.6 ft²
(d) 7203.2 ft²

19. The lateral area of a cylinder is 270π in.² and its height is 15 in. Find the circumference of the cylinder. Round to the nearest tenth.

(a) 9 in.
(b) 18 in.
(c) 30 in.
(d) 1.2 in.

20. Which of the following is the lateral area of an oblique cylinder? Round to the nearest tenth.

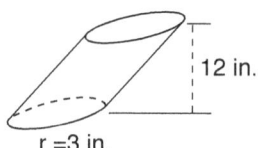

(a) 28.3 in.²
(b) 56.6 in.²
(c) 226.1 in²
(d) 452.2 in.²

21. If a cylinder's height is 22 cm and the diameter of its base is 6 cm, find the surface area of the cylinder. Round to the nearest tenth.

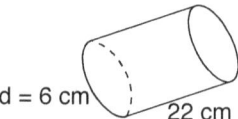

(a) 28.3 cm²
(b) 445.9 cm²
(c) 474.1 cm²
(d) 857.3 cm²

22. The circumference of the base of a cylinder is 36π m^2 with a height of 18 m. What is the lateral area of the cylinder? Round to the nearest tenth.

(a) 1196π m^2
(b) 1296π m^2
(c) 972π m^2
(d) 648π m^2

23. The surface area of a cylinder is 96 in.^{2} in the diameter of 12 in. What is the height of the cylinder?

(a) 5 in.
(b) 4 in.
(c) 3 in.
(d) 2 in.

24. The radius of a cylinder's base is 7 ft. Its height is 13 ft tall. What is the surface area of the cylinder?

(a) 140π ft^2
(b) 280π ft^2
(c) 232π ft^2
(d) 189π ft^2

25. Which of the following statements are false?

(a) The area of the base of a pyramid is B = lw, where l is the length and w is the width
(b) The surface area of a regular pyramid is the sum of the area of the base and the product of one half of the base perimeter and the slant height.
(c) The formula for the surface area of a regular pyramid is S.A. = B + $\frac{1}{2}$Pℓ.
(d) None of the above.

26. The rectangular pyramid has a slant height of 9 m, a width of 6 m, and a length of 4 m. What is the area of the base of the pyramid?

(a) 54 m^2
(b) 16 m^2
(c) 36 m^2
(d) 24 m^2

27. Which of the following is the lateral area of the square pyramid?

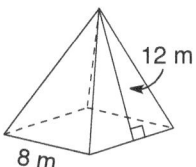

(a) 64 m^2
(b) 96 m^2
(c) 192 m^2
(d) 384 m^2

28. If a square pyramid has the lateral area of 168 ft^2 with a base length of 6 ft, find the slant height of the pyramid.

(a) 11 ft
(b) 12 ft
(c) 13 ft
(d) 14 ft

29. Find the area of the base of the pyramid. Round your answer to the nearest tenth if necessary.

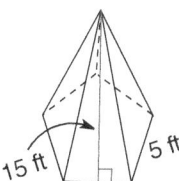

(a) 17.2 ft^2
(b) 34.4 ft^2
(c) 220.0 ft^2
(d) 229.4 ft^2

30. Use the diagram and the given information from Question **29**. Which of the following is the surface area of the pyramid?

(a) 150 ft²
(b) 177.5 ft²
(c) 92.2 ft²
(d) 184.4 ft²

31. Find the surface area of the prism. Round your answer to the nearest tenth if necessary.

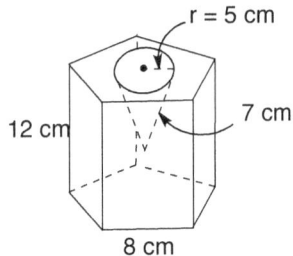

(a) 188.4 cm²
(b) 161.6 cm²
(c) 323.2 cm²
(d) 350.0 cm²

32. Given that a pyramid's slant height is 12 m and the length of its side is 2 m, find the apothem (*a*) of the pyramid.

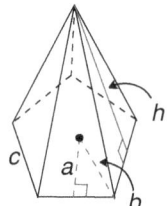

(a) $\dfrac{1}{\tan 36°}$ m
(b) $\dfrac{2}{\tan 36°}$ m
(c) $\dfrac{1}{\sin 36°}$ m
(d) $\dfrac{1}{\cos 36°}$ m

33. Use the diagram from Question **32**. If the slant height (*l*) is 9 ft and the side length is 4 ft, which of the following is the surface area of the pyramid?

(a) 89.8 ft²
(b) 359.0 ft²
(c) 179.5 ft²
(d) 207.1 ft²

34. Given a pyramid's slant height is 12 ft and the length if its base is 8 ft, find the surface area of the pyramid.

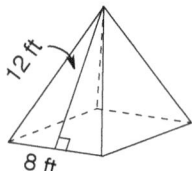

(a) 258 ft²
(b) 256 ft²
(c) 224 ft²
(d) 448 ft²

35. Which of the following is the base area of the pyramid?

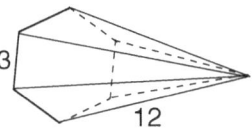

(a) 108 units²
(b) 27√3 units²
(c) 13.5√3 units²
(d) 216 units²

36. Use the diagram and given information from Question **35**. Find the surface area of the pyramid.

(a) 77.4 units²
(b) 154.8 units²
(c) 131.4 units²
(d) 108 units²

37. The lateral area of a cone is 439.6 ft². The diameter of the base is 14 ft. What is the height of the cone?

(a) 153.9 ft²
(b) 593.5 ft²
(c) 296.7 ft²
(d) 1055.0 ft²

38. If a right pyramid has a slant height of 25 cm with a base perimeter of 25 cm, what is the surface area of the pyramid?

(a) 351.6 cm²
(b) 703.1 cm²
(c) 312.5 cm²
(d) 337.5 cm²

39. If a right cone has a base diameter of 1 m and a height of 8 m, find the surface area of the cone. Round your answer to the nearest tenth.

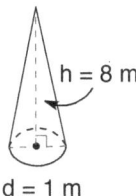

(a) 28.3 m²
(b) 4.3 m²
(c) 13.4 m²
(d) 9.0 m²

40. Which of the following statements are false?

(a) The surface area of a right cone is the sum of the area of the base and the lateral area.
(b) The lateral surface area of a cone is the product of the radius and the slant height.
(c) The formula of the surface area of a right cone is S.A. = $\pi r^2 + \pi r \ell$.
(d) None of the above.

41. Which of the following is the volume of the rectangular prism? Round your answer to the nearest whole number.

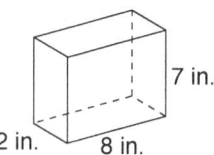

(a) 112.0 in.³
(b) 56.0 in.³
(c) 224.0 in.³
(d) 448.0 in.³

42. Which of the following is the volume of the right triangular prism? Round your answer to the nearest whole number.

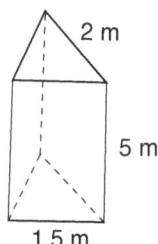

(a) 15.0 m³
(b) 7.5 m³
(c) 6.6 m³
(d) 13.1 m³

43. Which of the following is the volume of the right triangular prism? Round your answer to the nearest whole number.

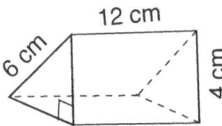

(a) 214.7 cm³
(b) 107.3 cm³
(c) 288 cm³
(d) 144 cm³

44. Which of the following statements are false?

(a) The volume of a prism is the product of the area of the base and the height.
(b) The formula of the volume of a prism is $V =$ Bh $= lwh$, where l is the length, w is the width, and h is the height.
(c) The volume of a cylinder is the product of the area of the base and the slant height.
(d) The formula of the volume of a cylinder is $V =$ Bh $= \pi r^2 h$, where r is the length, and h is the height.

45. Find the volume of the rectangular prism. Round your answer to the nearest tenth if necessary.

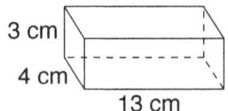

(a) 78 cm³
(b) 156 cm³
(c) 312 cm³
(d) 624 cm³

46. What is the volume of a rectangular prism with a height of 32 cm, a length of 18 cm, and a width of 6 cm?

(a) 6912 cm³
(b) 1536 cm³
(c) 1728 cm³
(d) 3456 cm³

47. What is the volume of a cube with a side length of 7 in.?
(a) 343 in.³
(b) 49.0 in.³
(c) 171.5 in.³
(d) 2401.0 in.³

48. What is the volume of the cylinder? Round your answer to the nearest tenth and give it in terms of π.

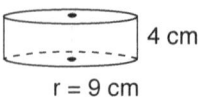

(a) 1296.0π cm³
(b) 432.0π cm³
(c) 648.0π cm³
(d) 324.0π cm³

49. What is the volume of the oblique cylinder? Round your answer to the nearest tenth and give it in terms of π.

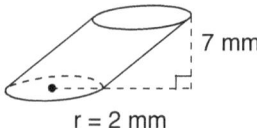

(a) 28.0 mm³
(b) 29.1π mm³
(c) 38.8π mm³
(d) 39.8π mm³

50. The height of the cylinder is 19 ft. The radius of the base of the cylinder is 6 ft. What is the volume of the cylinder? Round your answer to the nearest tenth and give it in terms of π.

(a) 684.0π ft³
(b) 342.0π ft³
(c) 2736π ft³
(d) 912.0π ft³

51. The volume of a cylinder is 943 ft³. The diameter of the base is 15 ft. What is the height of the cylinder? Round your answer to the nearest tenth.

(a) 16.8 ft
(b) 5.3 ft
(c) 4.2 ft
(d) 1.3 ft

Chapter 11 Surface Area and Volume of Solids

52. Which of the following statements are false?

(a) The volume of a pyramid is the product of one thirds of the area of the base and the slant height.
(b) The formula of the volume of a pyramid is $V = \frac{1}{3}Bh = \frac{1}{3}lwh$, where l is the length, w is the width, and h is the height.
(c) The volume of a cone is equal to one thirds of the volume of the cylinder.
(d) The formula of the volume of a cone is $V = \frac{1}{3}Bh = \frac{1}{3}\pi r^2 h$, where r is the length, and h is the height.

53. What is the volume of the cone? Round your answer to the nearest tenth and give it in terms of π.

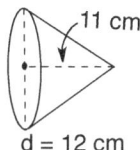

(a) 132.0π cm^3
(b) 44.0π cm^3
(c) 1584.0π cm^3
(d) 528.0π cm^3

54. What is the volume of the half cone? Round your answer to the nearest tenth and give it in terms of π.

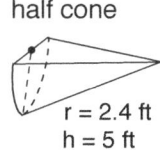

(a) 28.8π ft^3
(b) 19.2π ft^3
(c) 4.8π ft^3
(d) 9.6π ft^3

55. Find the volume of the right square pyramid. Round your answer to the nearest tenth.

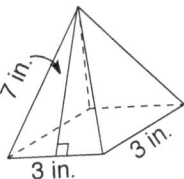

(a) 61.5 in.^{3}
(b) 20.5 in.^{3}
(c) 21.0 in.^{3}
(d) 63.0 in.^{3}

56. A height of a cone is 19 ft. The diameter of the cone is 6 ft. What is the volume of the cone?

(a) 228.0π ft^3
(b) 256.5π ft^3
(c) 171.0π ft^3
(d) 513.0π ft^3

57. If a height of a right pyramid is 28 m with a base area of 306.3 m^2, what is the volume of the pyramid?

(a) 11435.2 m^3
(b) 4288.2 m^3
(c) 2858.8 m^3
(d) 8576.4 m^3

58. Which of the following is the volume of the figure as shown below? Round your answer to the nearest tenth and give it in terms of π.

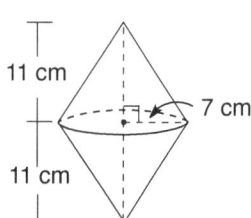

(a) 539π cm^3
(b) 179.7π cm^3
(c) 359.3π cm^3
(d) 107.8π cm^3

59. Which of the following statements are false?

(a) The formula of the surface area of a sphere is S.A. = $4\pi r^2$.
(b) The formula of the volume of a sphere is $V = \frac{4}{3}\pi r^3$.
(c) The surface area of a sphere is equal to one fourth of the area of a circle.
(d) None of the above.

60. What is the surface area of the sphere? Round your answer to the nearest tenth and give it in terms of π.

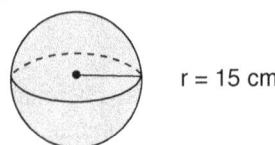

(a) 225.0π cm³
(b) 450.0π cm³
(c) 900.0π cm³
(d) 1125.0π cm³

61. What is the volume of the hemisphere? Round your answer to the nearest tenth and give it in terms of π.

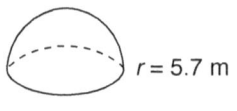
hemisphere

(a) 123.5π m³
(b) 246.9π m³
(c) 370.0π m³
(d) 65.0π m³

62. Find the surface area of a basketball with a diameter of 18 in. Round your answer to the nearest tenth and give it in terms of π.

(a) 1296.0π in.³
(b) 648.0π in.³
(c) 324.0π in.³
(d) 162.0π in.³

63. If the diameter of the sphere is 6 ft, what is the volume of the sphere? Round your answer to the nearest tenth and give it in terms of π.

(a) 18.0π ft³
(b) 36.0π ft³
(c) 108.0π ft³
(d) 54.0π ft³

ANSWERS

(1) b	(2) a	(3) b	(4) d	(5) c	(6) b
(7) a	(8) b	(9) c	(10) a	(11) d	(12) c
(13) b	(14) d	(15) b	(16) d	(17) c	(18) b
(19) b	(20) c	(21) c	(22) d	(23) d	(24) b
(25) d	(26) d	(27) c	(28) d	(29) a	(30) c
(31) b	(32) a	(33) c	(34) b	(35) c	(36) c
(37) b	(38) a	(39) c	(40) d	(41) a	(42) c
(43) b	(44) c	(45) b	(46) d	(47) a	(48) d
(49) a	(50) a	(51) b	(52) a	(53) a	(54) c
(55) b	(56) c	(57) c	(58) c	(59) d	(60) c
(61) b	(62) c	(63) b			

REVIEW TEST II

Chapter 11

1. Given that a polyhedron has 10 vertices and 15 edges, find the number of faces.

   (a) 5
   (b) 6
   (c) 7
   (d) 8

2. Find the lateral and surface areas of the rectangular prism. Round your answer to the nearest tenth if necessary.

   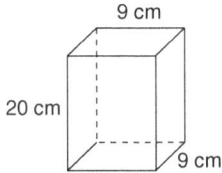

   (a) L.A. = 1440.0 cm^2, S. A. = 2961.0 cm^2
   (b) L.A. = 720.0 cm^2, S. A. = 1521.0 cm^2
   (c) L.A. = 1440.0 cm^2, S. A. = 1521.0 cm^2
   (d) L.A. = 720.0 cm^2, S. A. = 2961.0 cm^2

3. Find the height of a triangular prism that has a base perimeter of 28 ft and the lateral area of 140 ft^2.

   (a) 2.5
   (b) 5
   (c) 7.5
   (d) 10

4. A right cone has a base diameter of 8 cm and a slant height of 8 cm. Find the surface area of the cone. Round your answer to the nearest tenth.

   (a) 137.3 cm^2
   (b) 301.4 cm^2
   (c) 150.7 cm^2
   (d) 43.7 cm^2

5. Find the surface area of the regular pentagon to the nearest tenth.

   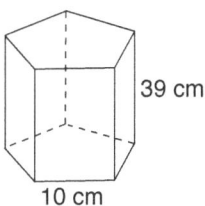

   (a) 172.0 cm^2
   (b) 344.1 cm^2
   (c) 2200.0 cm^2
   (d) 2294.1 cm^2

6. Find the surface area. Round your answer to the nearest tenth.

   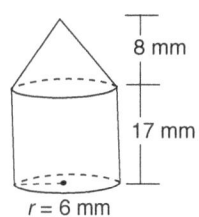

   (a) 1168.1 mm^2
   (b) 866.6 mm^2
   (c) 1130.4 mm^2
   (d) 429.0 mm^2

7. A lateral area of an oblique cylinder is 480 ft^2 with the radius 8 ft. Find the height of the cylinder. Round your answer to the nearest tenth.

   (a) 9.6 ft
   (b) 19.2 ft
   (c) 30.0 ft
   (d) 2.3 ft

8. Given that a cone's height is 4 in. and its diameter is 14 in., find the volume of the cone. Round to the nearest tenth.

(a) 196.0 in.^{3}
(b) 615.4 in.^{3}
(c) 157.9 in.^{3}
(d) 2461.8 in.^{3}

9. Find the volume of the oblique cylinder. Round your answer to the nearest tenth.

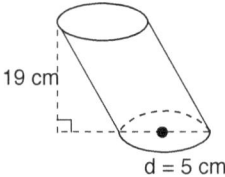

(a) 118.8 cm^3
(b) 475.0 cm^3
(c) 372.9 cm^3
(d) 1491.5 cm^3

10. Find the surface area of a right pyramid that has a base length of 15 m and a slant height of 21 m.

(a) 225 m^2
(b) 630 m^2
(c) 855 m^2
(d) 1485 m^2

11. Find the volume of the right square pyramid. Round your answer to the nearest whole number.

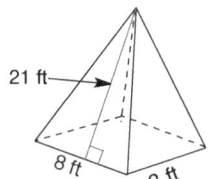

(a) 220 ft^3
(b) 440 ft^3
(c) 660 ft^3
(d) 1319 ft^3

12. A rectangular pyramid is 18 m tall. The length of the base is 24 m long and the volume of the pyramid is 3312 m^3. What is the length of its width?

(a) 24 m
(b) 23 m
(c) 22 m
(d) 21 m

13. What is the volume of the sphere? Round your answer to the nearest tenth and give it in terms of π.

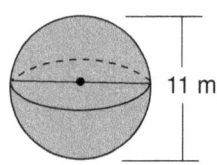

(a) 379.9π m^3
(b) 484.0π m^3
(c) 998.3π m^3
(d) 124.8π m^3

14. Find the volume. Round your answer to the nearest tenth and give it in terms of π.

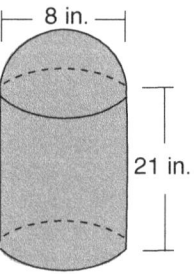

(a) 384.0π in.^{3}
(b) 848.0π in.^{3}
(c) 1392.0π in.^{3}
(d) 1728.0π in.^{3}

15. Which of the following is the ratio of the surface areas of the figures below?

(I) 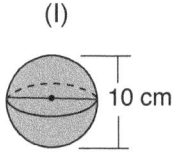 (II)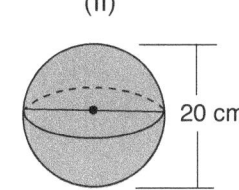

(a) 1 : 2
(b) 1 : 4
(c) 1 : 6
(d) 1 : 8

CHAPTER 10

16. Find the value of x.

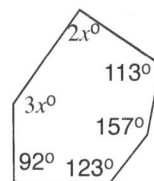

(a) 42
(b) 47
(c) 84
(d) 141

17. Use the given diagram to find the length of $\widehat{AB}$. Round your answer to the nearest tenth.

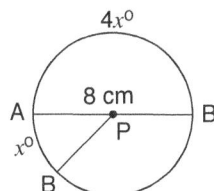

(a) 1.6 cm
(b) 3.2 cm
(c) 4.2 cm
(d) 6.3 cm

18. In the diagram, the radius of the circle is 7 cm. Given that ABCD is a square, find the area of the unshaded region.

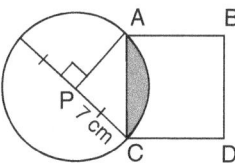

(a) 25 cm^2
(b) 38 cm^2
(c) 84 cm^2
(d) 98 cm^2

19. Which of the following is the area of the shaded region in the diagram?

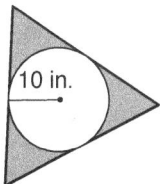

(a) 286 in.^{2}
(b) 314 in.^{2}
(c) 519.6 in.^{2}
(d) 1039.2 in.^{2}

20. The sum of the measures of the exterior angles of a convex pentagon is 360°. Which of the following is the value of x?

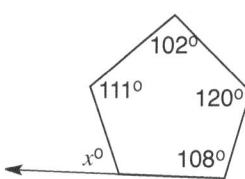

(a) 81°
(b) 99°
(c) 101°
(d) 110°

21. The measure of the given arc angle is 68° and the length of the radius is 3 in. What is the area of the sector of the circle below? Round your answer to the nearest tenth.

(a) 1.8 in.^{2}
(b) 3.6 in.^{2}
(c) 5.3 in.^{2}
(d) 6.1 in.^{2}

22. Find the areas of the shaded and unshaded regions of the regular polygon below. Round your answer to the nearest tenth.

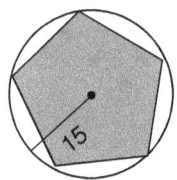

(a) A_{shaded} = 535.0 units^2, $A_{unshaded}$ = 706.5 units
(b) A_{shaded} = 388.7 units^2, $A_{unshaded}$ = 317.8 units^2
(c) A_{shaded} = 267.5 units^2, $A_{unshaded}$ = 439.0 units^2
(d) A_{shaded} = 535.0 units^2, $A_{unshaded}$ = 171.5 units^2

23. Find the value of x. Round your answer to the nearest tenth.

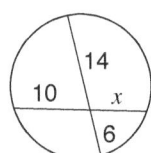

(a) 2
(b) 5.5
(c) 8.4
(d) 10

24. Find the measure of ∠C.

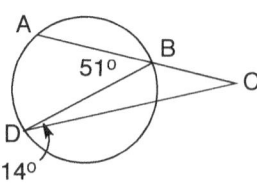

(a) 23°
(b) 28°
(c) 37°
(d) 74°

25. Find the perimeter (P) and area (A) of the square below. Round your answer to the nearest tenth.

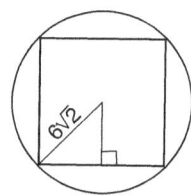

(a) P = 48 units, A = 114 units^2
(b) P = 24 units, A = 114 units^2
(c) P = 48 units, A = 203.6 units^2
(d) P = 24 units, A = 203.6 units^2

26. Find the length of $\widehat{AB}$.

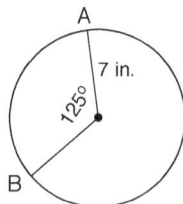

(a) 6.1 in.
(b) 7.6 in.
(c) 12.3 in.
(d) 15.3 in.

27. Find the circumference of circle P. If necessary, round to the nearest whole number.

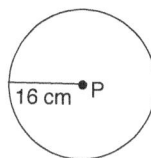

(a) 50 cm^2
(b) 100 cm^2
(c) 804 cm^2
(d) 1600 cm^2

28. Find the unshaded area of the circle below. If necessary, round your answer to the nearest tenth.

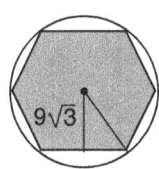

(a) 175.6 units^2
(b) 763.0 units^2
(c) 841.8 units^2
(d) 1017.4 units^2

29. Find the length of $\widehat{AB}$.

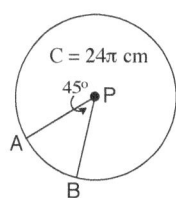

(a) 9.42 cm
(b) 18.84 cm
(c) 3.00 cm
(d) 12.00 cm

30. Use the diagram to find the area of the pentagon. Round to the nearest tenth.

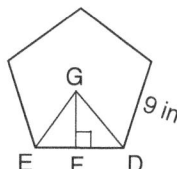

(a) 101.3 in.^{2}
(b) 139.4 in.^{2}
(c) 250.3 in.^{2}
(d) 344.5 in.^{2}

CHAPTER 9

31. Which of the following is the value of r in circle P?

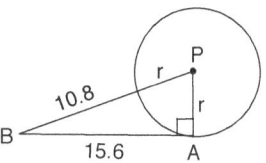

(a) 5.2
(b) 5.9
(c) 6.4
(d) 7.2

32. Given that the length of AD equals the length of DB, CE is 8 units and BE is 5 units, find the length of BD. If necessary, round your answer to the nearest tenth.

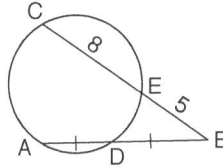

(a) 4.5
(b) 6.3
(c) 8.1
(d) 5.7

33. The measure of inscribed angle Q in circle P is 42°. Find $m\widehat{SR}$.

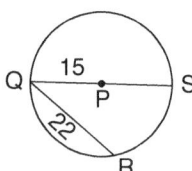

(a) 42°
(b) 96°
(c) 84°
(d) 48°

34. In the given information, $m\widehat{AB} = 115°$, and $m\widehat{AC} = 85°$. Find the value of x.

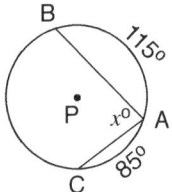

(a) 80°
(b) 115°
(c) 160°
(d) 150°

35. The length of AC equals the length of BC. Find the value of x.

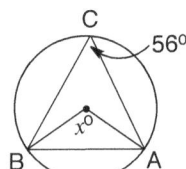

(a) 120°
(b) 118°
(c) 114°
(d) 112°

36. Use the diagram to find the value of x if the secants in the exterior of a circle are equal.

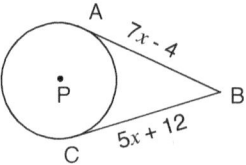

(a) 6
(b) 7
(c) 8
(d) 9

37. In the diagram, find the measure of ∠C if $m\angle FED = 20°$ and $m\angle BAF = 26°$.

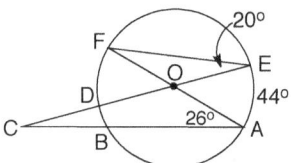

(a) 16°
(b) 20°
(c) 26°
(d) 18°

38. CD is a secant segment and AB is a tangent segment that shares the same endpoint A outside of circle P. If CD = 9 cm and AC = 7 cm, find the length of AB.

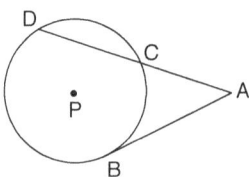

(a) 10.6 cm
(b) 7.9 cm
(c) 12.0 cm
(d) 9.8 cm

39. Find the value of x.

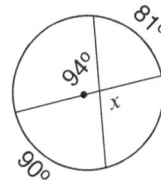

(a) 90°
(b) 92°
(c) 94°
(d) 110°

40. Find the length of AB. If necessary, round your answer to the nearest tenth.

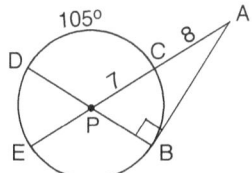

(a) 13.3
(b) 7.5
(c) 9.3
(d) 12.0

41. In the diagram, the lengths of AD and BD in circle P are congruent. If the length of AB is 20 cm and the length of PD is 7 cm, what is the length of the radius PC?

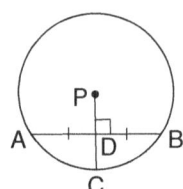

(a) 10.5 cm
(b) 12.2 cm
(c) 13.4 cm
(d) 14.0 cm

42. Two secant segments CE and AD share the same endpoint outside the circle. AD and DB are congruent segments. If the product of the length of CB and the length of BE is 65 cm, find the length of DB.

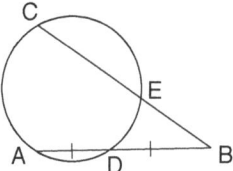

(a) 5.7 cm
(b) 5.5 cm
(c) 5.3 cm
(d) 5.1 cm

43. AB is tangent to the circle P. The measure of $\widehat{BD}$ is 125°. Find the measure of ∠A.

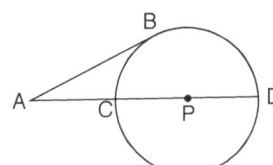

(a) 25°
(b) 30°
(c) 35°
(d) 42°

44. Find the coordinates of the center and radius of the circle given that the circle's standard equation is $(x - 5)^2 + (y + 2)^2 = 2$.

(a) (−5, 2) and $r = \sqrt{2}$
(b) (−5, 2) and $r = 2$
(c) (5, −2) and $r = \sqrt{2}$
(d) (5, −2) and $r = 2$

45. Write the standard equation of a circle using the given information. (notations: center = c and radius = r) c = (−2, 4) and r = 6

(a) $(x − 2)^2 + (y + 4)^2 = 36$
(b) $(x − 2)^2 + (y + 4)^2 = 6$
(c) $(x + 2)^2 + (y − 4)^2 = 6$
(d) $(x + 2)^2 + (y − 4)^2 = 36$

46. Find the quadrant of the coordinates of the center of a circle with the given equation. $(x − 2)^2 + (y − 5)^2 = 4$

(a) Quadrant I
(b) Quadrant II
(c) Quadrant III
(d) Quadrant IV

47. Which of the following is the equation of a circle with its center at P(−4, −2) and a diameter of 16?

(a) $(x + 4)^2 + (y + 2)^2 = 256$
(b) $(x + 4)^2 + (y + 2)^2 = 64$
(c) $(x − 4)^2 + (y − 2)^2 = 64$
(d) $(x − 4)^2 + (y − 2)^2 = 4$

48. What are the coordinates of the center of the circle as indicated by the equation $(x + 16)^2 + (y − 16)^2 = 16$?

(a) (4, −4)
(b) (16, −16)
(c) (−4, 4)
(d) (−16, 16)

CHAPTER 8

49. Use the given information in the diagram to find the value of x.

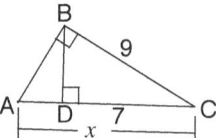

(a) 12.2
(b) 11.6
(c) 11.2
(d) 10.8

50. Find the value of x.

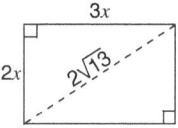

(a) 1
(b) 2
(c) 3
(d) 4

51. ΔABD and ΔCBD are similar right triangles that are separated by the altitude BD. The length of AD is four times longer than DC. If the length of the altitude BD is 8, find the length of DC.

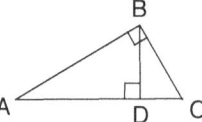

(a) 3
(b) 4
(c) 5
(d) 6

52. Use the given information in the diagram below to find the length of CD.

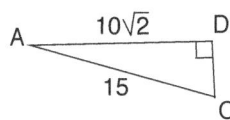

(a) 3
(b) 4
(c) 5
(d) 6

53. Find the measure of ∠A. Round to the nearest whole number.

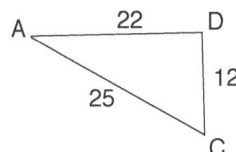

(a) 61°
(b) 42°
(c) 38°
(d) 28°

54. Find the value of x. Round to the nearest tenth.

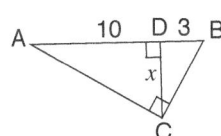

(a) 4.8
(b) 5.2
(c) 5.5
(d) 5.8

55. Find the values of x and y.

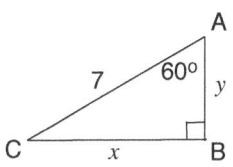

(a) $x = 3.5\sqrt{2}$, $y = 3.5$
(b) $x = 3.5\sqrt{3}$, $y = 3.5$
(c) $x = 4\sqrt{2}$, $y = 3.5$
(d) $x = 5\sqrt{3}$, $y = 4$

56. Find the measure of ∠A. Round to the nearest whole number.

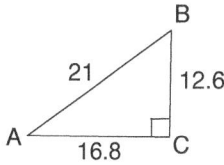

(a) 61°
(b) 58°
(c) 53°
(d) 37°

57. Find the value of x.

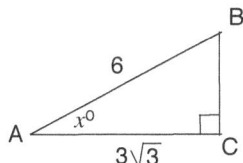

(a) 40°
(b) 35°
(c) 30°
(d) 25°

58. Find the values of x and y. Round to the nearest tenth.

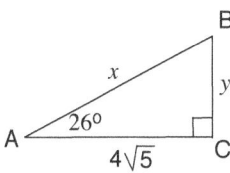

(a) $x = 10.0$, $y = 4.4$
(b) $x = 15.5$, $y = 4.0$
(c) $x = 18.3$, $y = 15.9$
(d) $x = 9.8$, $y = 4.0$

59. Use the given information to find the measure of ∠U. Round to the nearest whole number.

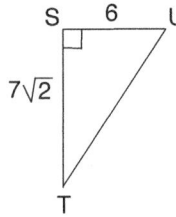

(a) 31.2°
(b) 52.7°
(c) 58.8°
(d) 60.4°

60. Find the measure of ∠A. Round to the nearest whole number.

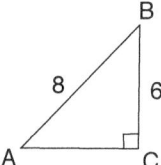

(a) 38°
(b) 41°
(c) 45°
(d) 49°

61. Classify the triangle as right, acute, obtuse, or neither.

$$10, 7\sqrt{3}, 6\sqrt{4}$$

(a) right triangle
(b) acute triangle
(c) obtuse triangle
(d) neither

62. Use compass directions to find the distance and direction of the vector.

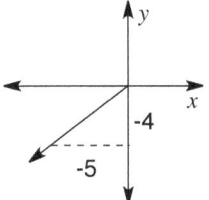

(a) 219°
(b) 224°
(c) 232°
(d) 245°

CHAPTER 7

63. The following polygons are similar. Find the value of y.

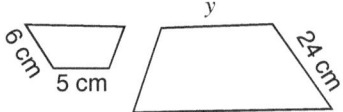

(a) 10
(b) 14
(c) 17
(d) 20

64. Find the value of x. If necessary, round to the nearest tenth.

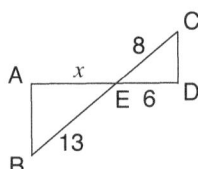

(a) 9.0
(b) 9.4
(c) 9.8
(d) 10.2

65. Find the value of x. If necessary, round to the nearest tenth.

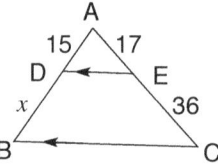

(a) 30.4
(b) 31.8
(c) 32
(d) 34

66. △ABC is similar to △DEF. Find the perimeter of △DEF if the perimeter of △ABC is 59.8 in. Round to the nearest tenth.

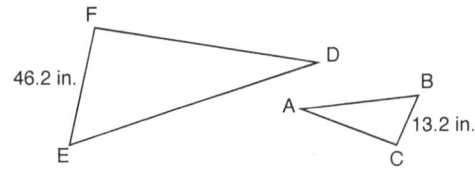

(a) 125.4 in.
(b) 178.5 in.
(c) 209.3 in.
(d) 789.4 in.

67. △ABC is similar to △ADE below. Find the length of DE. If necessary, round to the nearest tenth.

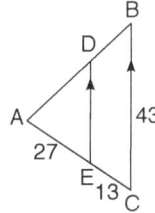

(a) 29.0 units
(b) 20.7 units
(c) 27.0 units
(d) 31.1 units

68. Find the value of x. If necessary, round your answer to the nearest tenth.

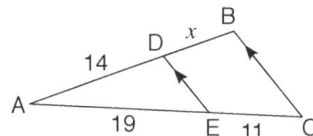

(a) 2.1
(b) 8.5
(c) 8.1
(d) 9.5

69. Find the value of x.

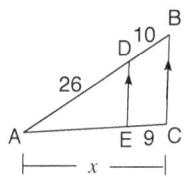

(a) 29.8
(b) 34.2
(c) 36.4
(d) 32.4

70. △ABC has vertices A (1, 3), B(3, 3), and C(3, 1). Which of the following coordinates are the vertices after the triangle is dilated given that the scale factor is 3?

(a) A'(3, 9)
(b) B'(9, 9)
(c) C'(9, 3)
(d) All of the above

ANSWERS

(1) c	(2) b	(3) b	(4) c	(5) d	(6) d
(7) a	(8) b	(9) c	(10) c	(11) b	(12) b
(13) d	(14) a	(15) d	(16) b	(17) b	(18) c
(19) a	(20) a	(21) c	(22) d	(23) c	(24) c
(25) a	(26) d	(27) c	(28) c	(29) a	(30) b
(31) b	(32) d	(33) c	(34) a	(35) d	(36) c
(37) a	(38) a	(39) b	(40) a	(41) b	(42) a
(43) c	(44) c	(45) d	(46) a	(47) b	(48) d
(49) b	(50) d	(51) b	(52) c	(53) d	(54) c
(55) b	(56) c	(57) c	(58) a	(59) c	(60) c
(61) b	(62) a	(63) d	(64) c	(65) b	(66) c
(67) a	(68) c	(69) d	(70) d		

APPENDIX

1. Distance and Midpoint Formulas and the Pythagorean Theorem.

The Distance Formula
$$AB = \sqrt{(x_2 - x_1)^2 + (y_2 - y_1)^2}$$

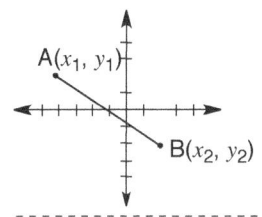

The Midpoint Formula
$$m_{AB} = \left(\frac{x_1 + x_2}{2}, \frac{y_1 + y_2}{2}\right)$$

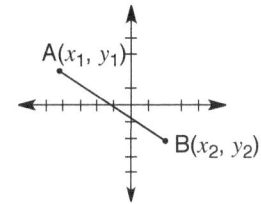

Pythagorean Theorem
$$c^2 = a^2 + b^2$$

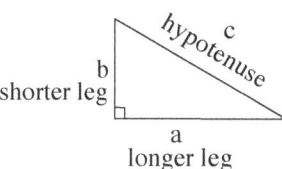

2. Useful formulas for areas of the polygons, and circumference of a circle.

Area of a rectangle

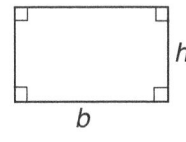

$A = bh$

Area of a parallelogram

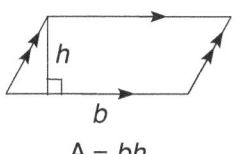

$A = bh$

Area of a triangle

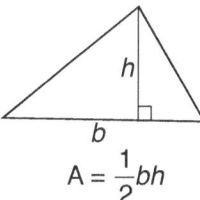

$A = \frac{1}{2}bh$

Area of a trapezoid

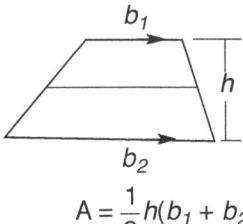

$A = \frac{1}{2}h(b_1 + b_2)$

Area of a kite

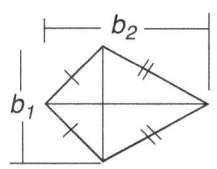

$A = \frac{1}{2}b_1b_2$

Area of a rhombus

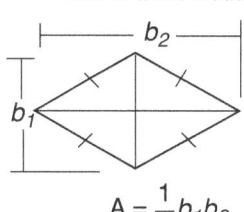

$A = \frac{1}{2}b_1b_2$

Area of an isotrapezoid

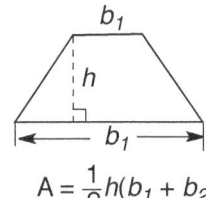

$A = \frac{1}{2}h(b_1 + b_2)$

Area of a square

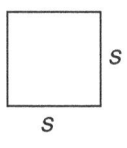

$A = s^2$

Area (A) and circumference (C) of a circle

$C = 2\pi r$

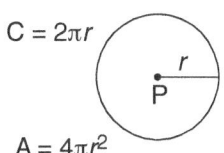

$A = 4\pi r^2$

3. Useful formulas for finding the lateral, base, and surface areas of solids.

Name of Solid	Lateral Area (L.A.)		Surface Area (S.A.)	
Right Prism	L.A. = ph	p = perimeter h = height	S.A. = L.A. + 2B S.A. = $ph + 2(lw)$	B = area of a base B = lw l = length w = width
Cylinder	L.A. = Ch or L.A. = $2\pi rh$ $C = 2\pi r$	C = circumference, h = height, r = radius	S.A. = L.A. + 2B S.A. = $2\pi rh + 2\pi r^2$	B = πr^2
Pyramid	L.A. = $\frac{1}{2}p\ell$ $\ell = \sqrt{a^2 + b^2}$	ℓ = slant height a = short leg b = long leg	S.A. = L.A. + B S.A. = $\frac{1}{2}p\ell + \frac{1}{2}ap$	Regular pyramid B = $\frac{1}{2}ap$ a = apothem p = perimeter
Cone	L.A. = $(\pi r \ell)$ $\ell = \sqrt{a^2 + b^2}$	ℓ = slant height	S.A. = L.A. + B S.A. = $\pi r \ell + \pi r^2$	B = πr^2
Sphere			S.A. = $4\pi r^2$	

4. Useful formulas for finding the volume of solids.

Name of Solid	Volume (V)	
Right Prism	V = Bh B = lw	B = area of the base, h = height, l = length, and w = width
Cylinder	V = Bh V = $\pi r^2 h$	B = πr^2 B = area of the base h = height
Pyramid	V = $\frac{1}{3}Bh$	B = lw B = area of the base h = height
Cone	V = $\frac{1}{3}Bh$, or V = $\frac{1}{3}\pi r^2 h$	B = πr^2
Sphere	V = $\frac{4}{3}\pi r^3$	

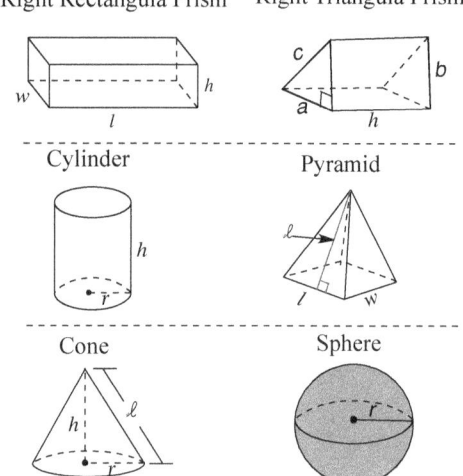

Right Rectangula Prism   Right Triangula Prism

Cylinder   Pyramid

Cone   Sphere

5. Useful formulas for the angles of polygons.

Name of Polygon	Formula (Angles, Areas, or Circumference)	
Polygon Interior Angles	n-gon = $(n - 2)(180°)$	n = number of side
Polygon Exterior Angles	n-gon = $\frac{1}{n}(360°)$	n = number of side
Area of an Equilateral Triangles	A = $\frac{1}{4}\sqrt{3}(s^2)$	s = side length
Area of a Regular Polygon	A = $\frac{1}{2}aP$ or A = $\frac{1}{2}a(ns)$	a = apothem, P = perimeter, s = side length, and n = number of side
Circumference of a Circle	C = πd or C = $2\pi r$	r = radius and d = diameter
Area of a Circle	A = πr^2	r = radius
Area of a Sector	$\dfrac{\text{Arc length of } \widehat{AB}}{2\pi r} = \dfrac{m\widehat{AB}}{360°}$	r = radius $m\widehat{AB}$ = measure of arc length

6. Useful information about isosceles triangles.

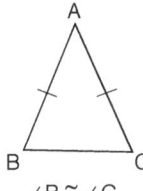
∠B ≅ ∠C

If *two sides* of a triangle are congruent, then *the angles* opposite those sides are congruent.

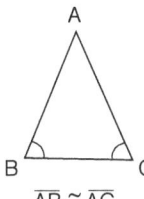
$\overline{AB} \cong \overline{AC}$

If *two angles* of a triangle are congruent, then *the sides* opposite the angle are congruent.

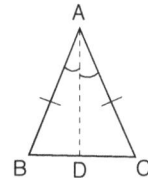
$\overline{AD} \perp \overline{BC}$ and $\overline{AD}$ bisects $\overline{BC}$

The bisector of the vertex angle of an isosceles triangle is *the perpendicular bisector* of the base.

7. Equiangular-Equilateral triangles and the Hypotenuse Congruence Theorem

Equiangular-Equilateral Triangle

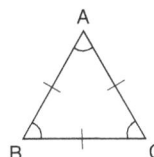

If a triangle is *equilateral*, then it is *equiangular*.
If a triangle is *equiangular*, then it is *equilateral*.

Hypotenuse-Leg(HL) Congruence Theorem

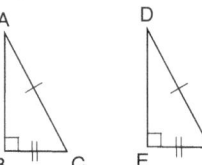

If $\overline{BC} \cong \overline{EF}$ and $\overline{AC} \cong \overline{DF}$, then △ABC ≅ △DEF.

If *the hypotenuse and a leg* of a right triangle are congruent to the hypotenuse and a leg of a second right triangle, then *the two triangles* are congruent.

8. Useful information for finding the lateral and surface areas of each prism.

Right triangular prism

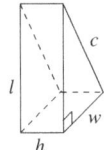

B = (1/2)(l·w)
L.A. = ph = (l + w + c)h
S.A. = 2B + L.A.

Equilateral triangular prism

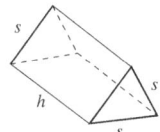

B = (1/4)(√3)(s²)
L.A. = ph = (s + s + s)h
S.A. = 2B + L.A.

Regular triangular prism

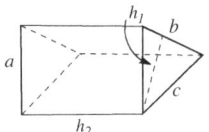

B = (1/2)(b)(h₁)
L.A. = ph = (a + b + c)h₂
S.A. = 2B + L.A.

Appendix

9. Useful information for finding the apothem of different polygons.

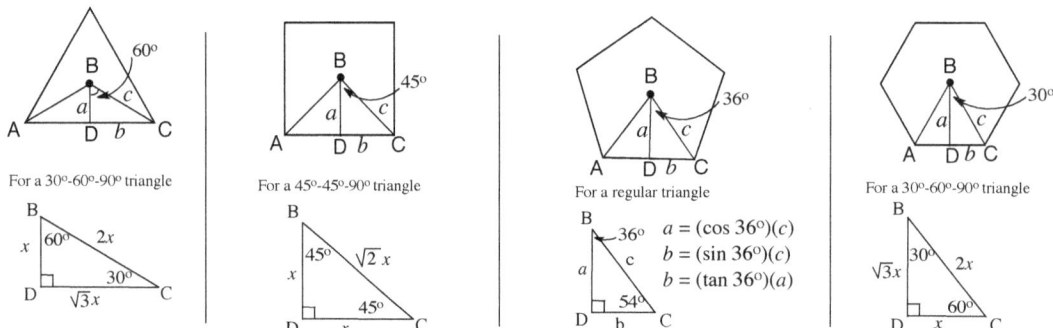

10. Vocabulary

Vocabulary	Definations/Characteristics	Diagram	Symbols
Point	No dimension and represented by a small dot.	• A	point A
Line	No endpoints and extends forever in two directions.	←•—•→ A B	$\overleftrightarrow{AB}$
Line segment	A part of a line that is bounded by two distict end points.	•—• A B	$\overline{AB}$
Ray	One endpoint and extends forever in one direction.	•—•→ A B	$\overrightarrow{AB}$
Plane	Extends forever in all directions.	A • C / B	plane AB

   Book (Part One)
* A polygon is a 2-dimensional closed shape that is made up of straight lines. (page 167)
* Acute angle: measures greater than 0°, but less than 90°. (page 5)
* Acute triangle: All three angles of a triangle are less than 90°. (page 97)
* Alternate exterior angles: If two parallel lines are cut by a transversal, then the pair of alternate exterior angles are congruent. (page 60)
* Alternate interior angles: If two parallel lines are cut by a transversal, then the pair of alternate interior angles are congruent. (page 60)
* Collinear: points on the same line. (page 1)
* Complementary angles: Two angles that add up to 90°. (page 7)
* Concave polygon: At least one angle in a polygon is more than 180°. (page 167)
* Conclusion: the "then" part of a conditional statement. (page 33)
* Conditional Statements: a type of logical statements that has two parts, a hypothesis and a conclusion. (page 33)
* Conjecture: A conjecture is a prediction that is based on observations. (page 2)
* Consecutive interior angles: If two parallel lines are cut by a transversal, then the pair of consecutive interior angles are supplementary. (page 60)
* Convex polygon: A polygon with all its interior angles less than 180°. (page 167)
* Coordinate: a set of numbers that corresponds to a point on a graph. (page 4)
* Coplanar: points on the same plane. (page 1)
* Corresponding angles: If two parallel lines are cut by a transversal, then the pair of corresponding angles are congruent. (page 60)
* Counterexample: A counterexample is an example that proves a conjecture to be false. (page 2)

* Equiangular triangle: All three angles (60°) are congruent. (page 97)
* Equilateral triangle: All three sides are congruent. (page 97)
* Exterior angle: An angle formed by any side of a polygon and the extended line of the adjacent side. (page 97)
* Hypothesis: the "if" part of a conditional statement. (page 33)
* Inductive Reasoning: Inductive reasoning is a process that involves looking for patterns and making conjectures. (page 2)
* Interior angle: The inside angles of a polygon. (page 97)
* Isosceles triangle: At least two angles and sides are congruent. (page 97)
* Kite: A quadrilateral that has two pairs of consecutive congruent sides. (page 174)
* Obtuse angle: measures greater than 90°, but less than 180°. (page 5)
* Obtuse triangle: One angle of a triangle is greater than 90°. (page 97)
* Paragraph proof: a type of proof written in paragraph form. (page 38)
* Parallel lines: two lines that never intersect each other. (page 59)
* Parallel planes: two planes that do not intersect. (page 59)
* Perimeter: The sum of the lengths of all the sides of the polygon. (page 9)
* Right angle: 90°. (page 5)
* Right triangle: One angle of a triangle is 90°. (page 97)
* Scalene triangle: All three angles and sides are NOT congruent. (page 97)
* Skew lines: lines that do not intersect and noncoplanar. (page 59)
* Supplementary angles: Two angles that add up to 180°. (page 7)
* The measure of each interior angle of a regular polygon is $\frac{(n-2)180°}{n}$, where n represents the number of sides of the polygon. (page 167)
* The sum of interior angle of any polygon is $(n-2)180°$, where n represents the number of sides of the polygon. (page 167)
* The sum of the measures of the exterior angles of any polygon is 360°. (page 167)
* Transversal: A transversal is a line that intersects two coplanar lines at two distinct points. (page 60)
* Two-column proof: a type of proof written as numbered statements and reasons that show the logical order of an argument. (page 38)
    * Vertical angles: Two nonadjacent angles are vertical from each other by two intersecting lines. (page 7)

Book (Part Two)
* Altitude: The segment from a vertex of a triangle perpendicular to the line containing the opposite sides. (part two--page 35)
* Chord: a segment whose endpoints are on a circle. (part two--page 69)
* Dilation: Dilation is an image that produced by enlarging or reducing a size, but not its shape. (part two--page 11)
* Equivalent ratios: two ratios that reduced to the same number. (part two--page 1)
* Proportion: a proportion is an equation that states that two ratios or fractions are the same. (part two--page 1)
* Ratio: a ratio is a comparison of two numbers by division. (part two--page 1)
* Secant: a line that intersects a circle in two points. (part two--page 69)
* Tangent: a line that intersects a circle in exactly one point. (part two--page 69)

Visit us at WWW.IQMATHS.com

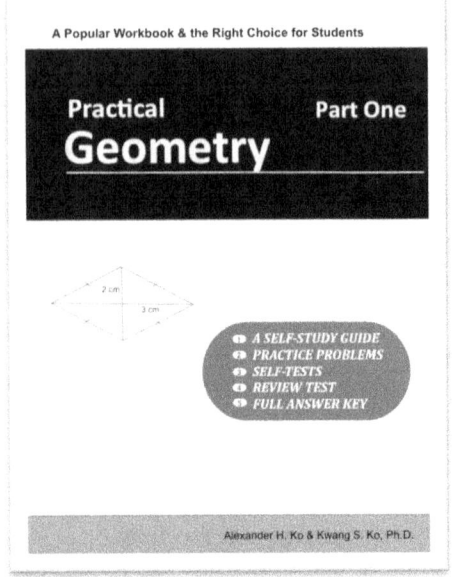

ISBN: 978-1-5232673-6-1

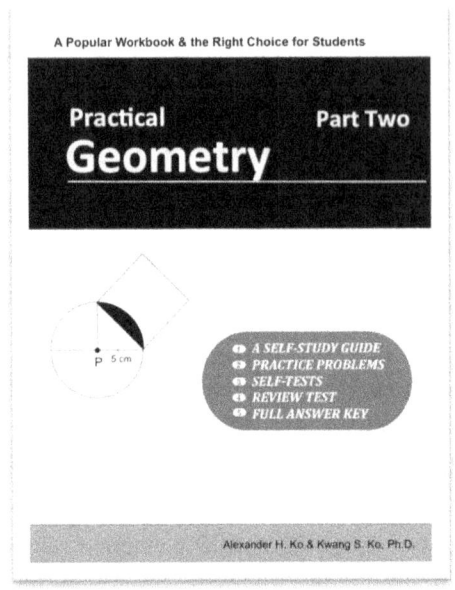

ISBN: 978-1-5233620-1-1

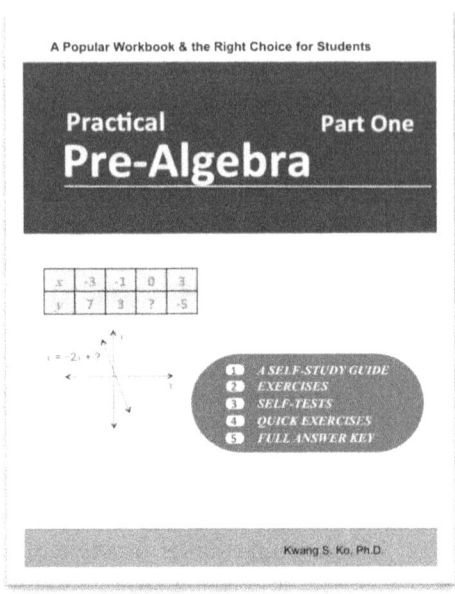

ISBN: 978-1-5233628-6-8

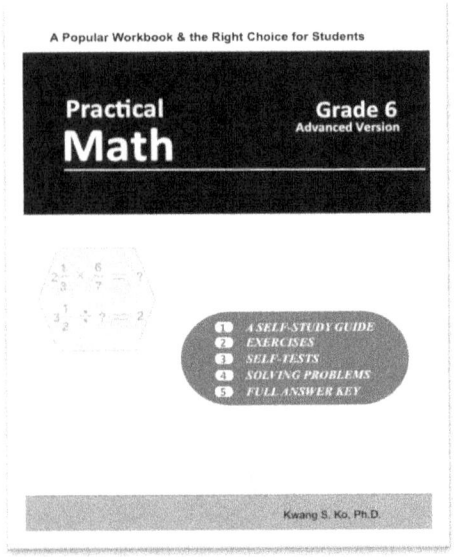

ISBN: 978-1-5233628-9-9

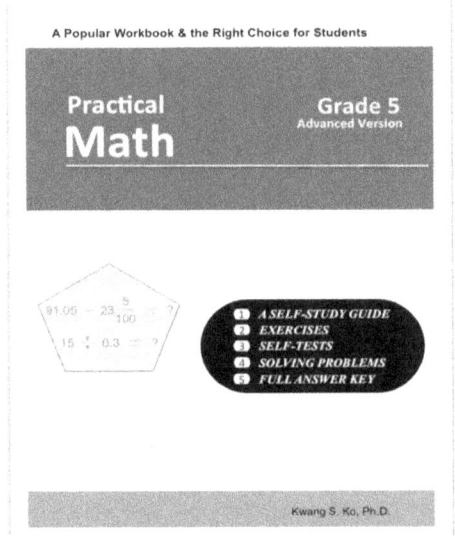

ISBN: 978-1-5233630-1-8

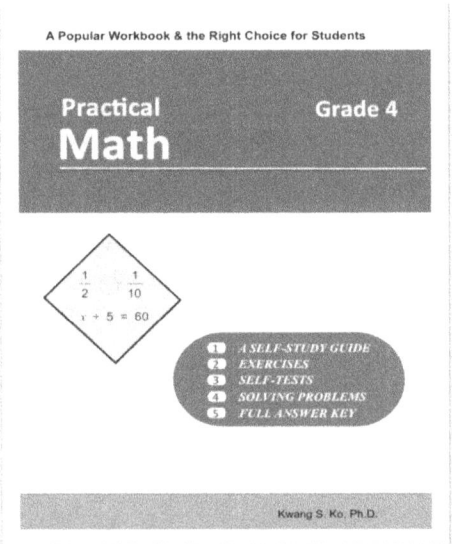

ISBN: 978-1-5233630-2-5

Other books are sold at WWW.IQMATHS.com.